AF743482

Greening Citizenship

Greening Citizenship

Sustainable Development, the State and Ideology

Andy Scerri

palgrave
macmillan

© Andy Scerri 2012

All rights reserved. No reproduction, copy or transmission of this publication may be made without written permission.

No portion of this publication may be reproduced, copied or transmitted save with written permission or in accordance with the provisions of the Copyright, Designs and Patents Act 1988, or under the terms of any licence permitting limited copying issued by the Copyright Licensing Agency, Saffron House, 6–10 Kirby Street, London EC1N 8TS.

Any person who does any unauthorized act in relation to this publication may be liable to criminal prosecution and civil claims for damages.

The author has asserted his right to be identified as the author of this work in accordance with the Copyright, Designs and Patents Act 1988.

First published 2012 by
PALGRAVE MACMILLAN

Palgrave Macmillan in the UK is an imprint of Macmillan Publishers Limited, registered in England, company number 785998, of Houndmills, Basingstoke, Hampshire RG21 6XS.

Palgrave Macmillan in the US is a division of St Martin's Press LLC, 175 Fifth Avenue, New York, NY 10010.

Palgrave Macmillan is the global academic imprint of the above companies and has companies and representatives throughout the world.

Palgrave® and Macmillan® are registered trademarks in the United States, the United Kingdom, Europe and other countries

ISBN: 978–1–137–01030–8

This book is printed on paper suitable for recycling and made from fully managed and sustained forest sources. Logging, pulping and manufacturing processes are expected to conform to the environmental regulations of the country of origin.

A catalogue record for this book is available from the British Library.

A catalog record for this book is available from the Library of Congress.

10 9 8 7 6 5 4 3 2 1
21 20 19 18 17 16 15 14 13 12

Printed and bound in Great Britain by
CPI Antony Rowe, Chippenham and Eastbourne

For Avel and Theo, and in memory of Sam

Contents

Tables

Preface: From Dualist Subjection of Nature to Holistic Participation in Nature

In recent decades, observers of Western society have sought to understand a significant transformation of citizenship as both a status and a practice. Prior to the 1950s, 'citizenship' had been conceptualized as the social institution that defines the status of individuals in relation to the state and other social institutions, especially markets. Interest in the concept had largely waned since then, when T.H. Marshall extended this basic definition to describe citizenship as not merely an 'official' category defined by the state but also as the practical, historical product of social conditions.[1] The more recent research builds upon and extends Marshall's work, yet draws attention to the emergence in the 1980s and 1990s of a 'green' transformation of citizenship that was also affecting its relationship to the state. For Bart van Steenbergen, Bryan S. Turner and others writing in these decades, a key impetus for this new 'greening' of citizenship was increased public awareness of what Ulrich Beck described as proliferating 'risks'.[2] In particular, what van Steenbergen and Turner note is that public antipathy towards the unquantifiable 'risky' consequences of techno-scientific decisions, such as on nuclear power or genetic technology, were coinciding with a significant decline in support for mass-political forms of organization and increased valuation of local community life, as well as a deepening of desires for individual autonomy. For van Steenbergen in particular, this shift had brought into contention, while also fundamentally altering, the status of citizenship in relation to the state. Meanwhile, for Turner, the new citizenship was altering how discourses of justice and injustice were being enunciated, bringing into contention what Marshall had described as the 'social' form of citizenship that had been established through the institutional arrangements of the welfare state in the 1950s and 1960s. Amidst the relatively cosmopolitan and articulate culture of the postindustrial Western states, Turner saw citizens' lack of confidence in expert opinion and lack of confidence in government policy for dealing with risk as affecting the embeddedness of citizenship within the state itself and, in turn, 'eroding' state capacities to administer the rights and duties of social citizenship. This is because the civil, political and social rights and duties that had once been largely satisfied by state-based institutions – representative government, courts of law and the social welfare apparatus – were by the 1990s being overburdened by new demands for global human and in some cases non-human rights for both present and future generations.[3]

What is noticeable in van Steenbergen's and Turner's analyses is that this political erosion of support for social citizenship and the welfare state also has a cultural dimension. That is, concerns with the unintended 'environmental' consequences of social activities, relative lack of interest in mass-political organizations, such as parties and trades' unions, and the embrace of relatively individualistic, 'life-enhancing' inter-relations at the human scale of local communities can be understood in cultural terms as a partial consequence of the normalizing of new social and countercultural movement values, which had spread across the postindustrial West since the 1960s and 1970s. Whereas in the early 1970s Charles Reich had argued that a counterculture-led *Greening of America* was imminent, by the 1980s and 1990s it had become clear that these once marginal cultural values had been integrated into both progressive and reactionary political positions. Seen as a transformation akin to what some postmodern theorists argued at the time was a generalized cultural incredulity towards grand historical narratives – of progress, of scientific fact or of subjugating nature and controlling wayward individuals and populations – the new social and countercultural movements challenged both 'the Establishment' and progressive social movements, such as social democratic parties and trades' unions. From the 1980s onwards, the greening of citizenship that van Steenbergen and others describe called into question the prevailing cultural grammar that had hitherto enframed political concerns with justice. Whether defined in reactionary terms as a problem of distributing wealth to those deserving it or in progressive terms as a problem of distributing wealth to all equally, the new citizenship challenged the focal point for debates over justice, extending it beyond the issue of distributing wealth within the state. Born of wider cultural and political change, this 'greening' or 'erosion' of traditional social citizenship confronted the welfare state compromise between the state, organized labour and industrial capitalism as a barrier to universalizing human equality, freedom and solidarity and, especially, opportunities for self-realization, authenticity, peaceable relations with others and harmony within nature.

Adopting an 'ideological' perspective, van Steenbergen regards the political and cultural greening of citizenship as a symptom of declining support for what had long been the prevalent *dualistic* cultural ideological view of society as rightfully engaged in a collective effort to completely dominate or subdue nature. In place of such ideological dualism, the greening of citizenship serves to highlight what he defines as growing support for a *holistic* cultural ideological view of society as an active participant in nature.[4] Whereas the cultural ideological dualism that was long associated with modernization had represented the social whole in terms of the sum of its fundamental parts, the introduction of holism through the greening of citizenship meant that social relations were increasingly being framed within a grammar that defined the social whole as a complex system, the functioning of which could not acceptably be understood in terms of its component parts only. That is, the greening of citizenship seems to represent a social shift and an ideological transformation, away from the dualism that had enframed debates over justice in terms of a grammar that represented the main social problem as one of dividing the economic spoils created by society, rightfully organized around the collective task of subduing nature. With the greening of citizenship, such debates were being reframed in terms of a shared view of the main social problem as one of how society should interact within the ecosphere.

Extending this research, in the 1990s there arose a series of normative theoretical interventions that, building upon the observations of van Steenbergen, Turner, Beck and others, seek to describe what a 'holistic' society of green citizens should look like. These theories provide normative arguments for what should be the types of social and political participation, and the rights and duties and the institutional arrangements of green citizens *qua* the state or, more ambitiously, some form of post-state polity. The normative theories respond to what the early observers identify as the erosion of social citizenship by linking green citizenship with broader and more abstract normative ideals describing rights and duties, the basis for social relations, notions of political space and, importantly, justice. For the normative theorists, the greening of citizenship has meant that justice becomes a problem of defining the terms for 'sustainable development'. Regardless of whether the greening of citizenship refers to the emergence of gradualist market-driven eco-modernization or deep-green wholesale change in social life, both have the objective of re-defining the good society as the sustainable society. From within the perspective of green citizenship, justice becomes not merely a matter of redistributing wealth amongst citizens within a political community but of how society's participation in nature is to be organized globally to address relations between humans of present and future generations, that is, over the inherently uncertain short, medium and long terms.

In this sense, questions once couched solely in economic-redistributive terms become matters of what Tim Hayward regards as 'the fundamental human right to an environment adequate for [individuals'] health and wellbeing'.[5] This succinct statement defines a 'green' ideal of justice by evoking a holistic narrative of human belonging within the ecosphere, rather than a dualistic narrative that regards justice as the product of equitably distributing the 'spoils' derived from society's unquestioned effort to dominate and subdue nature. However, it also offers a means for moving beyond extensive and abstract normative debate about the coherence of different subcategories of holism or dualism, which would emphasise one or the other as the more rational basis for debating justice in the context of unsustainable development. The most widely debated form of holism – ecocentrism – defends the view that nature itself has intrinsic value, while dualism – most often defined as anthropocentrism – defends the view that human society is the source of all value. The interesting thing about Hayward's assertion in this respect is that although it begins from a 'weak' anthropocentric premise – that respect for Nature can and should be grounded in respect for the Self and Others[6] – his point seems to be holistic, insofar as justice and sustainable development are together and necessarily regarded as the by-products of collective efforts to live fairly within the capacity of the ecosphere, rather than of collective efforts to 'master' nature as an unlimited cornucopia.

A central contention developed in Part I of this book is that the greening of citizenship – when regarded as a consequence of a cultural ideological shift from dualism to holism that has significant political ramifications, rather than as a normative project concerned to elaborate what the ideological and practical dimensions of green citizenship should be – does not of itself imply that holism entails a progressive form of justice. Taking it that ecocentrism is but one expression of ideological holism, I find that counteracting nature/culture dualism has not directly fostered justice. This is because holism may not be a political but rather a cultural ideological issue, in that it provides the shared representational grammar or encompassing context within which citizens evaluate and deploy different political discourses as contributions to social life. I argue that the normalizing of holism has created new opportunities for progressive movements oriented to achieving justice, just as it has fostered opportunities to define justice in reactionary terms of preserving existing structures of privilege. Hence, I do not discuss differences between anthropocentric 'environmental' and ecocentric 'ecological' green citizens; rather, I draw attention to how changing social conditions have over time shaped the greening of citizenship, and then examine these as the setting for an ideological frame for action within which progressive, or reactionary, ideas of justice may be elaborated in the context of unsustainable development.

As an admittedly somewhat provocative alternative to the normative theoretical approaches, which regard green citizenship as defining the ethical parameters of what the green good life should be and the moral parameters of how green citizens should treat each other, I find that the greening of citizenship has not been an entirely good thing. I develop an approach that allows it to be understood as the source of both bottlenecks and opportunities for progressive social movements acting to achieve justice. Throughout the book, I develop the idea that since the publication of van Steenbergen's 1994 essay, the political impacts of the greening of citizenship have been somewhat underestimated. In particular, I find that what van Steenbergen, Turner and others observed as widespread recognition that society participates in nature, rather than acts upon it, needs in the early 21st century to be taken as a definitive ideological transformation. I find that the normalizing of cultural ideological holism has impacted how progressive political movements must work to realize justice. In short, I argue that with the greening of citizenship, progressive assertions that injustice is present have become more complex, while reactionary ideas about justice have narrowed. In a holistic frame, reactionary claims that justice is served through the pursuit of narrow self-interest no longer need to appeal to secondary concerns with upholding tradition, particularism over universalism and essentialism over contingency. These merely refer to ever-greater negative freedoms that will further unburden 'deserving' individuals and communities in their narrow pursuit of security from risk through untrammelled consumerism. Meanwhile, I find that the progressive political position is bound to support institutions that, as David Schlosberg argues, redistribute rights fairly, recognize as equal all social participants and provide transparent representation or effective participation while delineating the territorial reach of the political community of citizens against global aspirations to remain within the capacity of the ecosphere to provide for present and future generations.[7] And, following Nancy Fraser, I find that in this sense 'it is not only the substance of justice, but also the frame, that is in dispute',[8] such that progressive assertions that injustice is present require domestic and international support for policy that enables capabilities for human flourishing at the subjective level and equality of capacities to contribute, at the societal level, to global efforts to remain within the constraints set by the ecosphere.[9]

In Part II of the book, chapters develop the argument that achieving progressive forms of justice and, so, sustainable development implies competing normative premises and evaluative standards, and various formulations, identifying different human causes of social and environmental problems and prescribing different remedies for those problems,[10] remedies that need to be accepted by citizens as actors within a political community that is more than likely defined by, if not commensurate with, the state. That is, sustainable development cannot just be about technical decisions or, indeed, upholding nature's inherent value, but must be the product of political and cultural decisions to support certain rights and responsibilities, and the practical and institutional arrangements that come with them. 'Remedies' to the problem may extend from relatively 'weak' options, supported by the administratively led adjustment of policy tools that might uncritically accommodate deep contradictions, to the 'strong' transformation of political and economic institutions, which is necessarily the product of citizenly critique of social conditions and relations.[11] That is, change tends to be forced upon the state by citizens, while also being enacted by the state on behalf of citizens. This is because states remain the key sources of legitimate political authority to organize the rules for life held in common amongst citizens who, as well as being political actors, share a cultural narrative that describes their belonging within the ecosphere and amongst each other. For these reasons, I regard securing a progressive form of justice as a problem that is inexorably linked with cultural ideas about what counts as a meaningful narrative of social belonging within the ecosphere *and* with political challenges to the *status quo* that seek to effect how the rules for life held in common are organized and, by extension, how these impact others. By recognizing the greening of citizenship as a partially successful rather than a yet-to-be implemented project, I conclude by examining how some contemporary progressive social movements are responding to the continuing reproduction and 'export' of injustice by postindustrial ecomodernizing Western states.

A key impetus in developing this argument is recent work by Luc Boltanski and Eve Chiapello, Fraser and Axel Honneth. In Boltanski and Chiapello's argument, a novel form of exploitation has been fashioned from the new social and countercultural movement critique of bureaucratic welfare statism, such that much of what was once emancipatory in relation to industrial society now motivates and supports socially and environmentally destructive postindustrial processes of accumulation.[12] Moreover, these theorists all notice what Fraser concedes is the recent 'disturbing convergence' of the new movement ideals 'with the demands of an emerging new form of capitalism – post-Fordist, "disorganized", transnational... neoliberal'.[13] Fraser deals specifically with feminism, but argues that the utopian desires of the New Left currents that supported it in the 1970s and 1980s have, more recently, found a second life as feeling currents that unwittingly help to legitimate postindustrial injustices. Meanwhile, for Honneth, the individualization of demands for self-realization once tabled by critical new social movements have morphed, and currently support the displacement of social infrastructure by neoliberal markets that demand ever greater self-responsibility in relation to matters hitherto dealt with collectively.[14]

Throughout the book, I argue for a similar view of the greening of citizenship by regarding it as having coalesced around five central normative claims: the need to challenge nature/culture dualism; to dissolve the divide between the public and private spheres; to eschew social contractualism; to undermine territorialism; and to ground justice in awareness of finite and maldistributed ecological footprint or 'ecospace'. The problem that I address is that, from within the postindustrial ecomodernizing 'global competition state',[15] the injustices associated with unsustainable development appear to be framed by holism, the lack of a clear distinction between ethico-moral and political obligations, the dissolution of the social contract, globalized deterritorialization and widespread awareness of finite ecospace. That is, the central claims of the normative theories of green citizenship might just be *for another time*. The central critiques seem to challenge a dualistic, Fordist, industrial, state-centric and conformist society that no longer exists.

As a Westerner born in the final year of the 1960s, and so old enough to have experienced the decline of the industrial solidarity that sustained welfare statism and the metamorphosis of counter- into consumer culture, and to have witnessed the battle to save the Tasmanian Franklin River and unfolding media reports of the disaster at Chernobyl, all in the context of rising social inequality and deepening ecological crisis, I find the line of questioning that Boltanski and Chiapello, Honneth and Fraser open up to be deeply interesting. To paraphrase Fraser, the book asks on what terms certain aspirations that had a clearly emancipatory thrust in the context of industrial society have come to assume a far more ambiguous meaning in 21st-century postindustrial conditions. I trace a heuristic narrative that links the political and cultural developments that Boltanski and Chiapello, Fraser and Honneth observe with the greening of citizenship, the state and ideology. Through it, I try and understand how many of the political hopes held by greens have, in recent decades, been enlisted in the service of achievements that run contrary to a larger vision for strong sustainable development.

Acknowledgements

Writing this book has been an adventure. And, like any adventure, it has had its lows and highs. I am deeply thankful to those who offered help, support and encouragement as I worked through the research and writing process. I thank Paul James and Manfred Steger of the Global Cities Research Institute and Martin Mulligan and Damian Grenfell of the Globalism Research Centre, as well as Tom Nairn, Kim Humphery and Heikki Patomäki, all currently or formerly at RMIT University. I also thank my colleagues Anne McNevin, Erin Wilson, Liam Magee, Rob Cameron, Selver Sahin, Tommaso Durante and Victoria Stead, as well as Alan Roberts, Arran Gare, Bobby Roberts, James Oliver, Sophie Bibrowska and Vivienne Waller for their support and advice. I also thank John Barry, Andrew Biro, Anthony Elliott and Bryan S. Turner, as well as the publisher's anonymous reviewers, for their helpful comments on my work. I am grateful to Andrew Gorman-Murray and Ruth Lane for introducing me to the literature on social practice and consumption, and to Caroline Norma for helping me to think more clearly about the division between public and private spheres. I am especially indebted to Meg Holden for constantly raising the bar in relation to my interpretation of the French critical pragmatists and for introducing me to the literature on 'the world's most liveable cities', as well as for what became highly enjoyable annual visiting scholar placements in the Urban Studies Programme at Simon Fraser University. I thank Hayley Stevenson and the other members of the Environmental Politics and Policy Standing Group of the Australian Political Science Association. And, Peter Christoff, Robyn Eckersley, David Schlosberg and Simon Tormey, who provided opportunities to develop seminar papers, and whose comments helped to clarify my thinking on questions of citizenship and justice towards the end of my work on the final chapters. I am thankful to James Goodman, who has long influenced my thinking on critical social movement theory, and to Victorian Commissioner for Environmental Sustainability, Kate Auty for challenging me to think more critically about public participation. I am deeply grateful for the support of Christina Brian, Senior Commissioning Editor, and Amanda McGrath, Editorial Assistant, at Palgrave Macmillan. Finally, I thank my dear friend Antonia Settle for her endless patience and uncommon sense in relation to both the academic and personal matters that completing this project raised. Of course, I stand responsible for all of the shortcomings of the book and its argument.

The work for the book was carried out in large part thanks to the atmosphere of intellectual inquiry and professional freedom that has been fostered over several years by Paul James, Manfred Steger and others at RMIT's Global Cities Research Institute. It would also not have been possible without the support of Martin Mulligan and Damian Grenfell, successive directors of the Globalism Research Centre, and David Hayward, Dean of the School of Global, Urban and Social Studies at RMIT. Elements of the work were funded by the Australian Research Council Discovery Grant DP077030, led by Tom Nairn, and Linkage Grant LP0990509, as well as by small grants made available by the Academy of the Social Sciences in Australia/Department of Innovation, Industry, Science and Research International Science Linkage Programme, the International Council for Canadian Studies Faculty Research Programme and the Canadian Government Social Science and Humanities Research Council Research Development Initiative. Elements of the argument presented in Chapters 5 and 6 have been substantially modified on the basis of research published in *Environmental Politics* 18(4), 2009: 467–85, *Management of Environment Quality* 21(10), 2010: 122–35 and the Springer Verlag *Encyclopaedia of Quality of Life* (2013).

Andy Scerri
Melbourne, June 2012

Part I

1
Introduction: Citizenship, the State and Ideology in a Critical, Pragmatic and Realist Lens

1.1 Normative theories of green citizenship

In some ways motivated by Andrew Dobson's claim that 'the form of citizens' daily lives – their "participation" in the widest sense – is what shapes the contours of sustainability itself',[1] the key normative theories of green citizenship that emerged in the 1990s and 2000s seek to inform policy and practice by outlining the kinds of rights and/or duties that the political community and individuals should support. These normative theories elaborate different conceptions of 'environmental', 'ecological', 'active' or 'critical sustainability' citizenship. As in any debate over a concept, there exists some disagreement among the proponents of green citizenship about its most important aspects. This said, such disagreement is the product of collegial effort to gain clarity and build the robustness of the concept. Primarily, these differences centre on what is the best balance between 'liberal' rights to something from society and 'civic-republican' duties and responsibilities to society. As in general political theory, critics of the liberal position argue that it fails to make clear what citizens must contribute in order to benefit from the polity. Meanwhile, critics of the civic-republican view regard it as limited to describing obligations, without making clear how claims to the shared benefits of belonging to the polity should be apportioned. Green liberal perspectives are said to over-emphasize entitlements; conversely, civic-republican perspectives are criticized for expressing a kind of green-tinged ethico-moral authoritarianism. Although in political philosophy the concepts of 'freedom' and 'rights' and 'duties' and 'responsibilities' stand apart, in practice this separation between what a citizen should be entitled to and what a citizen should be accountable for can, as Hayward suggests, be regarded as inseparable. Environmental (liberal, rights-based) and ecological (civic-republican, responsibilities-based) conceptions of citizenship are not in this view analytically distinct but share the same 'architecture' or, as Dobson contends, the same 'terrain', as examples of green citizenship.[2]

This said, on the one hand, Marcel Wissenburg recognizes that for liberal citizenship to become green, classical notions of negative liberty, (state) neutrality and anthropocentric bias can logically be limited without undermining the central notion of freedom, even though it does have the inevitable result that negative liberty can no longer be seen as the supreme criterion of a good society.[3] This greening of traditional liberalism is to be achieved through recourse to acceptance of biophysical ecological limits to neutrality and an ecological expansion of neutrality to encompass 'non-human and non-present human interests and the means for accounting for the formation of individual preferences'.[4] For Derek R. Bell, liberal green citizenship would necessitate the abandonment of classical liberal disembodied and territorial citizenship because it is incompatible with 'the right to have our basic needs met and the "fact of reasonable pluralism", as these must be understood in the context of humanity's desecration of the Earth's capacity to provide for it'.[5] In the main, these green forms of liberalism extend claims about civil, political and social rights to encompass environmental rights and also extend the definition of who is covered by such rights: if pollution is global, then it follows that citizenship should be thought about and acted on in global terms and should cover nonhumans. Hence, the liberal conception seems to offer a rights-based, and so contractual, perspective upon the assertion that state territorialism should be undermined. Pushing the liberal envelope further, however, Manuel Arias-Maldonado finds that the challenges of 'greening' liberalism while 'democratising democracy' along green lines 'are too deep to preserve the "liberal" label in all its force', and require that democracy itself be reconceived, 'not as a device for the aggregation of preferences and interests but as a dialogue within discursive communities'.[6] Based on the assertion that individuals' rights and responsibilities should be central to politics, liberal green citizenship overflows the boundaries of the nation-state, such that local community participation and a deliberative and consensus-driven politics represent the best means for establishing the ends of global sustainable development. Hence, the work by Wissenburg, Bell and Arias-Maldonado also has the important implication of bringing green liberalism closer to its green civic-republican cousin. Both dimensions seek to transcend territorialism and, to some extent, social contractualism, and normalize a local-global community-society. This is the case whether green citizenship would be grounded in civic-republican non-contractual 'virtuous' relations or non-reciprocal obligations to uphold justice in reference to ecospace *or* in liberal demands for deliberation within discursive communities that seek to settle the redistribution of rights and duties by debating these locally in view of global considerations.

In this light, and on the other hand, for John Barry the tendency for liberal green citizenship to privilege, however weakly, liberal 'contractual' rights over civic-republican conceptions of active, critical and responsible 'sustainability citizenship' is a problem. Barry argues that a more ambitious form of green citizenship, grounded by the task of 'connecting solidarity, commitment and democracy to citizenship', would challenge not merely the environmental but also the social dimensions of unsustainable development.[7] 'Ecological' civic-republican duties-based green citizenship seeks to cultivate citizenly virtues and an ecological identity, behaviour, values and, so, practices. Green citizenship in this aspect is promoted by political engagement and activism aimed at improving ecological awareness, scientific knowledge of climate change, self-reliance, self-restraint, self-responsibility and consideration of non-human interests. In this view, green citizenship is committed not merely to achieving environmental justice through formalized liberal rights but at the same time to achieving social justice by supporting citizens' active participation in decision-making aimed at achieving strong sustainable development.[8] From this claim alone, it may follow that a 'social contract' is unnecessary because citizens are the self-responsible contributors to or interlocutors within the political task of authorizing the rules for life held in common.

For Dobson, 'contract' is also an important issue that debates over the relationship between liberal and civic-republican concepts of green citizenship overlook, eliding the shortcomings of both because they are grounded in a shared 'social ontology'.[9] That is, both liberal and civic-republican theories presume contractualism:

> [such that citizenship] is regarded as a contract between the citizen and the state, in which the citizen claims rights against the state, but according to which the citizen also undertakes to contribute to the state's ends by paying taxes, for example, and by seeking work when unemployed. This is a reciprocity in which rights are earned.[10]

Dobson identifies, in the emergent post-cosmopolitanism that for him embodies contemporary citizenly demands for justice, signs of a cultural readiness for both gradualist environmental citizenship and more radical ecological citizenship, one that moves beyond both liberal and civic-republican conceptions. For him, an ideological commitment to contract does not so much define citizenship as '[associate] it closely with the juridical-economic sphere and the expectations and assumptions [of rational self-interested behaviour] that lie therein'.[11] Dobson alternatively 'suggests the possibility of *unreciprocated and unilateral* citizenship obligations' that are to be found in a post-cosmopolitanism that, as well as being non-contractual, is non-territorial and, as such, global while also being local, inhabiting 'a political space that [should] include the private as well as the public realm [and a] focus on virtue and its determination to countenance the possibility of "private" virtues being virtues of citizenship'.[12] In this way, green citizenship links private ethico-moral obligations or choices with public political obligations or actions. This dissolution of a division between the private and political spheres of social life is most commonly discussed in relation to citizens' personal ethico-moral choices to purchase or not particular commodities, and extends well beyond the literature on green citizenship itself. In theories of the citizen-consumer, in particular, as in the green citizenship literature, non-contractualism coalesces with the view that private virtues can and should be the virtues of citizenship.

In a recent critique of the green citizenship literature, Teena Gabrielson finds that such theories can implicitly describe a 'narrative of declension'.[13] For Gabrielson, traditional theories of citizenship define the 'good life' as one of political engagement in the public sphere *per se*, and from this vantage point look askance upon its displacement in modernity by inferior, privatized concerns with 'life alone'. Thus, citizenship theory itself harks back to some or other golden era of comprehensive and virtuous democratic participation, which it attempts to resuscitate. And green theories of citizenship tend similarly to decry the shift from a holistic relationship between humanity within the ecosphere and its displacement by modern nature/culture dualism. The theories of green citizenship that Gabrielson reviews hark back to a golden era of holistic and virtuous participation in nature. Indeed, understanding nature/culture dualism in these terms, its transformations over time and its refutations, has long been a key concern of environmental political and sociological theorizing, as well as of political ecology.[14] On the one hand, I agree with Gabrielson's observation that green citizenship can tend to dwell in a kind of elitist, 'enlightened' understanding of dualism as a condition of societal modernization and, furthermore, that such implicit declension-ism may grow out of a lack of empathy with those who for good reasons experience 'real nature' as something alienating. With Gabrielson, I agree that the virtues of agricultural life are felt most strongly when somewhat remote from everyday experience.[15] Normative arguments for green citizenship that aspire to replicating archaic relations within pristine nature or some kind of virtuous rural idyll not only have limited appeal but also seem to be products of an individualistic and urban social class fraction[16] that is forced, by consequence of circumstance, to act in ways that render 'private vices public virtues'. Formed in relative freedom from material necessity and the unsavoury aspects of keeping animals or working the soil, such a grouping 'looks back' on nature as the source of liberation from grey, polluted cities and atomized, alienated social relations. It is in this sense that Gabrielson finds that green citizenship fosters a kind of abstract utopianism, as do others, including Robert E. Goodin and John Dryzek and Patrick Dunleavy, who argue that highly normative theories of *Homo ecologicus* open greens to criticism as the purveyors of 'wishful thinking about how human sensibilities could be more benign towards nature'.[17]

On the other hand, while theories of green citizenship seem to argue from an implicit declension-ism, they also tend to build upon an explicitly normative rejection of nature/culture dualism in favour of holism, a move that Gabrielson and Katelyn Parady endorse. On this point I differ both with the normative theorists of green citizenship and with Gabrielson and Parady, who reject declension-ism yet embrace holism. For them, 'Few concepts are as deeply embedded in the dualisms of Western political thought as that of citizenship.' Following Charles Taylor, Gabrielson and Parady define dualism as 'the epistemological construal of modern thought [that] creates a disengaged subject "free and rational to the extent that he [*sic*] has fully distinguished himself from the natural and social worlds" so much so that "the subject withdraws from his own body, which he is able to look on as an object" '.[18] The authors recognize that the explicit aim of green citizenship is, following Dobson, to 'disrupt a set of binary oppositions' that are said to be central to citizenship in general and green citizenship in particular. These oppositions, between public and private, active and passive, rights and duties, and territorialized and de-territorialized space, are 'written into the very concept of citizenship' and, as such, should be challenged.[19] For Dobson, the basis upon which such dualism should be challenged is by operationalizing the central virtue of green citizenship: a commitment to justice that is grounded in the fair distribution of use of the ecological footprint, that is, one's ecospace.[20] In his view, awareness of the finiteness of ecospace – the total amount of biologically productive land and water area required to produce the resources consumed and to assimilate the wastes generated using prevailing technology – makes real for citizens their responsible role in reproducing the causal relations of unsustainable development and, so, calls upon them to exercise the 'always-already' private-public, de-territorialized, non-contractual and holistic obligations of green citizenship. In contrast, Barry contends that it is more so awareness of society's 'parasitic' exploitation of the ecosphere that should give rise to ethico-moral commitments to act responsibly and 'cultivat[e] ecologically virtuous modes of interaction with the environment' or green citizenly 'dispositions of character'.[21] For Dobson, however, Barry's view is purely an ethico-moral argument, 'appropriate to the Good Samaritan rather than to the Good Citizen'.[22]

Against such a clear line drawing together justice and responsibility for it, however, sits Hayward's theoretical objection to Dobson's argument. For Hayward, in the social context a moral community does not automatically create a political community, and moral obligations do not automatically translate into political obligations. Dobson argues that the normalizing of green citizenship implies widespread acceptance that an emergent conception of justice, as being grounded in citizens' ethico-moral recognition of the shared and limited nature of global ecospace, and the local-to-global causal relations that it brings to light, grounds political community and obligations.[23] For Hayward, the problem is that grounding justice in awareness of the local-to-global ecological implications of social reproductive processes does not necessarily foster political obligations. Even while the subjective realization that particular individuals or groups justifiably possess rights to only a 'fair share' in global ecospace – which may (and often does) foster personal reflection and prompt social movement action – it does not ground a political community sufficient to sustain *political* obligations. This issue is central to Hayward's engagement with Dobson. For Hayward, virtues such as embracing or promoting the reduction of one's use of ecospace do not automatically create a political community and, so, obligations.[24] What is required for translating ethico-moral into political obligations is a political sphere, be that of an ambiguously 'greening' postindustrial state or a post-state global order, either or both of which would hold the legitimate authority to organize the rules for life held in common around the distribution of responsibilities and rights by reference to ecospace. As such, Hayward's concern in pressing the link, or lack thereof, between ethico-moral and political community and ethico-moral and political obligation I take to be a problem that requires addressing.

In this book, I develop the view that the shortcomings of key areas of contemporary social and political practices, as these relate to justice, cannot be seen adequately from within the perspective provided by the five key claims central to normative theories of green citizenship (see Table 1.1). Hence, while I share the views of Wissenburg, Bell, Dobson, Barry and others that defining sustainable development by reference to ecospace is a desirable normative goal, I approach the topic in different terms from those used in the normative theories. The contention that I develop in practical terms in this book is that when operationalized through widespread contemporary practices and discourses, such as those of localized governance networks, the corporate social and environmental responsibility movement, 'green' consumerism or the proliferation of 'liveability indices', aspects of green citizenship's normative critique *are* being put into practice, yet ethico-moral obligations are not being translated into political obligations to uphold justice. Indeed, this is a practical danger that both Dobson and Barry recognize. For Dobson, green 'citizenship cannot only be a matter of individuals trying to reduce their own ecological space occupancy, where appropriate, but also involves those same individuals working in the traditional spaces of the public sphere to move society as a whole in a different direction'.[25] Dobson argues convincingly that while 'the management of change [to implement low-energy sustainable development] can't simply involve a return to the centralist corporatist politics of the 1970s, it can't rely on a thoroughgoing localism either, [as there exist] guarantees of collective behaviour that only government legislation can begin

Table 1.1 Central claims in normative theories of green citizenship

	The requirement
1.	to challenge nature/culture dualism;
2.	to dissolve the divide between the public and private spheres;
3.	to eschew social contractualism;
4.	to undermine (state) territorialism; and
5.	to ground justice in awareness of the finiteness and maldistribution of ecospace.

to provide'.[26] Barry also argues for a form of citizenship that takes the initiative and responsibility for 'ecological stewardship' and recognizes that the state must be the focal point or reference for action.[27]

What is missing from the discussions about breaking down dualism, bridging the private-public division, supporting non-contractual social relations, local-global non-territorialism and awareness of finite ecospace is an appreciation of how currently unsustainable societies promote holistic voluntarism, privatize political obligations, support social non-contractualism and de-territorialized social relations and reduce awareness of global ecospace to quality-of-life concerns *and* how this has few, no or even negative consequences for sustainable development. Moreover, I contend that, like the normative theories, many contemporary practices side-step the political institutions that, established over time as permutations of the state, are in place for transposing widely held ethico-moral obligations or virtues into political ones. By tracing out a narrative of change over time that focuses on the relationship between citizenship, the state and ideology amidst the ambiguous greening of the state, this book argues that the re-establishment of holism reveals possibilities *and* bottlenecks for elaborating progressive notions of justice. Against the backdrop of the postindustrial rolling-back of the welfare state, deepening of individualization and heightened awareness of risks, concern for these issues may encompass progressive forms of justice but may also support privatized means of securing against ill-health or self-interested responses to environmental hazards, for example. I discuss how some contemporary progressive movements, such as the environmental justice movement or those seeking to steer the profits generated by carbon emissions trading schemes towards benefiting those in need of support, the movement to regulate fair-trade relations within and between states and to promote 'science in society' might be understood as challenging injustice directly from within a political culture that has, for better and worse, come to see itself as a participant in nature.

To reiterate my central motivation, there is a need for inquiry into the opportunities and bottlenecks that the normalizing of ideological holism creates for achieving progress towards strong sustainable development. Indeed, as noted already, the merits or not of holism, most often cast in ecocentric terms, can provide the entry point to discussions of green political theory. Indeed, for Barry, the critique of holistic discourses of deep ecologism or ecocentrism actually now provides what 'seems like the obligatory' opening for green theoretical discussions.[28] While this book avoids these debates, it is interesting to recognize that Dobson has lately rethought the need to ground green politics in eschewing dualism, which he equates with ecocentrism, claiming that achieving green political ends is 'not about nature's intrinsic value but about the human species adapting to a long era of low-energy living'.[29] If Dobson means, by 'low-energy living', living within the capacity of the ecosphere as a precious and finite thing that is recognizably part of us as we are part of it, then he is expressing a holistic sentiment, one just as holistic as the idea that nature of itself emanates intrinsic value. For him, however, this shift 'wasn't [made] because ecocentrism (as an idea) didn't exist any more, but because environmentalists didn't seem to talk about it so much ... the sound of ecocentrism has been drowned out by the sound of pragmatic environmentalism'.[30] With this point in mind, I address the question of opposition to dualism in both practical and theoretical terms by developing my suspicion that what has 'drowned out the sound' of opposition to dualism might be the din that has been generated by its own partial successes. Through engagement with theoretical literature and a series of examples, I offer an explanatory account of how the eschewal of dualism has emerged as the *leitmotiv* for Western culture since the 1980s and 1990s. My point is that counter-dualism has failed to respond adequately to its position in relation to the historical development of Western politics because it is a cultural ideology. That is, what the early theorists of green citizenship saw as the emergence of a view of society as participating in nature has been insufficiently dealt with in terms of its partial success as an aspect of a broader cultural ideological 'turn' towards holism. An ambiguous and highly unsatisfactory greening of the state is taking place, yet is being overlooked because of the emphasis in normative theories upon dissolving nature/culture dualism, breaking down the split between private and public actions and implementing non-contractualism, non-territorialism and the ethico-moral dimensions of awareness of ecospace. This is a pressing issue if green political and social theory is to contribute to the further greening of the state or the establishment of a green global polity.

1.2 After dualism, beyond fatalism ...

The argument of this book is greatly motivated by two works on green political social theory. First, John M. Meyer argues for a new approach to contrasting *dualist* and *derivative* interpretations of the relationship between Western politics and nature. For him, the dualist interpretation holds that 'politics (and human culture generally) is completely divorced from nature', while the derivative interpretation 'derive[s] from conceptions of nature, whether that conception be the teleology of Aristotelians, the clocklike mechanism of early modern scientists, or the invisible hand of Darwinian selection'.[31] In both cases, 'nature is the inescapable *subject* of politics so conceived'. Meyer argues that a viable green politics must address the inadequacies of both dualist and derivative interpretive frames, because '*even if* dualism were dominant throughout [social history], overcoming it would not ensure the realization of environmental goals'.[32] My argument is based on the view that the real or imagined dominance of dualism *has* been overcome and that this has not, as Meyer rightly argues, brought about justice or sustainable development. Following Meyer, I contend that there are two main cultural expressions of society's approach to nature (dualist versus derivative), both of which require countering or support by progressive politics. Secondly, Andrew Biro offers a detailed examination of how the relationship between society and nature is represented in Frankfurt School critical theory. For Biro, in modernizing conditions the 'biological instinct for self-preservation',[33] hitherto constrained by an all-encompassing and explicit holism, is cut loose and individuals are effectively left to their own devices amidst a nature that emanates objective, positively knowable laws. As the world becomes the artifice of individual wills, binding holistic norms – including many of those associated with the inter-relational commonalities and continuities that are indispensable for social existence – begin to manifest as limitations on an autonomous and self-orienting individuality. This displaces older dominant senses of a binding totality that confirms subjectivity in relation to cosmologically ordained, unquestionable myth and tradition. In conditions of modernization, persons faced with adversity are set against an objectified universe of material forces, and must decide what to do *as if* for themselves. Biro concludes that a degree of such alienation from nature is necessary for critical politics and indeed social life. Following Biro, I contend that critiquing the terms on which society reproduces this alienation is the central challenge for progressive political movements.

In the light cast by these theorists' work, the debates around dualism might be said to spring from what Bruno Latour famously sees as the demise of a 'dualist illusion' that for a long time was premised upon the subjection of an external and objectified nature.[34] However, *pace* Latour, my argument is that while we may never have been modern, we did actively believe so for a long time and do so no longer. This recognition – that, illusion or not, dualism has passed – has become an established feature of the political and cultural landscape in the West, but has not taken place on terms of greens' choosing. With this recognition comes a responsibility to debate green political judgement and notions of justice. Hence, more recently, Latour has argued that it is the 'idea of nature' that divides what is objective and indisputable from what is subjective and disputable. For him, the solution to the politics of (society as participant in) nature lay in uniting the two – objective and subjective, human and non-human – realms by giving them an equal seat at the collective table of citizenship. This political collective would facilitate 'propositions' from all actors or actants, human or non-human, and prompt a return to civil peace that 'redefines politics as the progressive composition of a common good world'. Such a collective would accept actant 'propositions' as expressing the issues that should be considered if green ends are to be met. The political ecological collective would 'stir the entities of the political collective together... and make them speak'. This book develops an account of society that is partly commensurate with Latour's view that non-domination is a product of the 'collective' rather than two separate 'Houses' (of nature and society).[35] In part in response to Latour's claim that 'we have never been modern', my focus is upon the particular questions that are posed in relation to understanding the political culture of citizenship in the West *after* dualism.

Of course, a number of theorists employ systems-functionalist, realist and post-structuralist frameworks for developing 'post-dualist' accounts of contemporary environmental politics. I discuss a selection of these post-ecologist, post-environmentalist and post-political or post-democratic theories here, before setting out the approach that I develop for this book. However, while accepting some of the implications of their arguments, I hope to avoid what I see as their shared methodological fatalism.

Referring to Beck's well-known argument that a nascent form of green citizenship had arisen as a reaction to pervasive risk in the 1980s and 1990s, Ingolfur Blühdorn and Ian Welsh contend that the proliferation of ever-present and unquantifiable 'risks', as well as widespread awareness of these amongst a relatively well-educated populace, undermines public confidence in the authority of mass institutions: the state, the techno-sciences, major corporations and mass social movements. They argue that, at the same time, the predominance of the autonomous and, so, critical subject of modernity has been broken down. Using Niklas Luhmann's systems theory as an analytical framework, Blühdorn argues that this erosion of 'autonomy', which he regards as the defining feature of modernity, prompts a wide-reaching 'abdication of critique' on the part of Western individuals. For Blühdorn, contemporary citizens are no longer interested in developing a robust political critique of social conditions and relations. For Blühdorn and Welsh, 'the traditionally modern concept of the *autonomous subject* [that sustained critical social movements]...has run its course', such that contemporary society is shaped more by a pervasive cultural *'identification of the Self with the system*, i.e. the collapse of traditional modernity's central dualism of the *Self* and its *Other*'.[36]

The theoretical point is that the 'system' has encroached so totally upon the 'life-world' that critique from within the latter is no longer possible, nor do individual citizens desire it. The major normative claim is that citizens are said to have abandoned politics and macro-social institutions as they, ironically, have come to more readily accept the naturalness of structural social conditions. This shift is said to undermine the activist-driven ecologist paradigm that characterized Western political culture until the 1960s and 1970s, when the new social movements arose as an alternative to modern mass movements such as trades' unions. This transformation signals a transition from traditional to late, or what Beck, Anthony Giddens and Scott Lash define as 'reflexive modernity'. For Blühdorn, 'reflexive modernity' fosters a situation in which worldly-aware and self-aware individuals, including many environmentalists, actively take part in self-consciously ineffective symbolic protests or 'publicity stunts' that are designed not to challenge but, rather, implicitly to support unsustainable economic growth. Blühdorn and Welsh argue that green activism has been rendered impotent by the establishment of an 'opportunity society' and a culture that has come to be characterized by quietist 'post-materialism'. This situation gives rise to a 'post-ecological politics of unsustainability' and concomitant institutional orchestration of simulated action on what are nonetheless recognized as common social goods, such as sustainable development.[37] The best that can be hoped for amidst the post-ecological condition is that ostensibly green states will manage societal 'unwillingness to become sustainable' by adopting piecemeal reforms, such as inadequate voluntary energy conservation measures for individuals or ineffective self-regulatory industry schemes. That is, the decline of the radically autonomous subject implies the entrenching by *fiat* of very weak or neoliberal ecomodernization.

Interestingly, others advocate a similar view of politics – as having entered a post-environmentalist era – from within a more or less 'traditional' realist perspective. In their controversial 2004 work, *The Death of Environmentalism*, Michael Shellenberger and Ted Nordhaus argue that greens should move away from narrow-frame activism, based around technical solutions that promote the uptake of hybrid automobiles or energy-saving light globes, for example, and embrace a broad-frame approach that constructs the environmental problems of climate change or pollution as social problems. For Shellenberger and Nordhaus, 'Environmentalists are in a culture war... It's a war over our core values as Americans and over our vision for the future, and it won't be won by appealing to the rational consideration of our collective self-interest.' Moreover, for the authors, 'Protecting the environment is indeed supported by a large majority – *it's just not supported very strongly.*'[38] In this perspective, environmentalist activists are 'out of touch' with 'average people', insofar as the former subscribe to a Romantic or Utopian vision while the latter regard the ecological crisis as one amongst many socially created problems, the most important being that of ongoing individual, family and societal prosperity. Around this observation, Shellenberger and Nordhaus argue that popular support for green ends might be bolstered by using the strategies of mass-media advertising and promotional industries. These, having been very successful in shaping popular tastes for consumer goods and services, demand emulation in order to articulate green issues in ways that coincide with 'mainstream' values. More recently, Shellenberger and Nordhaus have expanded this approach to argue for a 'secular modernization theology' that in fact highlights and embraces technological progress as the 'social' solution to the ecological challenge. Rather than focusing on some ecological utopia or dystopia, the authors follow Latour in recommending 'a worldview that sees technology as humane and sacred, rather than inhumane and profane'. This subtle shift from eschewing technical solutions, as being narrow-focus symptoms of a mistaken 'environmentalist' value system, towards embracing technology *per se* as part of a broad-focus expression of some kind of sacred 'modernization theology' comes at a cost. That is, Shellenberger and Nordhaus' embrace of technology as the vehicle for expressing 'profound gratitude to Creation – human and nonhuman'[39] is important, insofar as it can serve to draw green critique away from knee-jerk anti-scientism. However, it may be the case that their enthusiasm for a secular religion of modernization elides possibilities for opening to debate questions concerning the social constitution of injustice in the production of an unsustainable development trajectory.

Finally, adopting a broadly post-structuralist approach, Eric Swyngedouw finds that enthusiasm for political engagement has diminished where post-political or post-democratic conditions, which brook no opposition, establish the growth-at-any-cost *raison d'être* of globalizing neoliberalism. Following Chantal Mouffe and Jacques Rancière, Swyngedouw characterizes the contemporary West as having entered a post-political condition that 'forestalls the articulation of divergent, conflicting and alternative trajectories of future socioenvironmental possibilities... solidifying a liberal capitalist order for which there seems no alternative... [and] which eludes conflict and evacuates the political field'.[40] Efforts to institute alternatives to the prevailing order are in this view merely elitist expressions of a 'populist fantasy'. However, despite describing how the establishment of the post-political system undermines possibilities for a genuine politics built around local participation and civic engagement, Swyngedouw calls for nothing less than an arch-radical green-democratic 'socioenvironmental political program [that will] crystallize around imagining new ways to organize processes of socio-metabolic transformation'.[41] This is a problem insofar as Swyngedouw seems to apply an approach to his topic that is simultaneously geared to recognizing a system that brooks no opposition and to recognizing alternatives to this otherwise totalizing system. The result is problematic, as it sets on a pedestal those privileged enough to hold insight into the 'populist fantasy' and to have 'imagination' sufficiently developed to re-organize 'processes of socio-metabolic transformation'. As such, Swyngedouw's approach risks replicating the mistakes of earlier political-theoretical vanguards.

The observations of Blühdorn and Welsh, Shellenberger and Nordhaus, and Swyngedouw make important contributions to debates over the greening of citizenship, the state and ideology *after* dualism. Interesting is Blühdorn and Welsh's call for a cool head of the kind that Luhmann advocates bringing to social inquiry:

> [What is required of inquiry is a] focus on detecting and investigating the strategies by which latemodern societies are trying to cope with the awareness and the apparent inescapability of their unsustainability and the full range of its ecological, social, cultural, political and economic consequences. The paradox of postecologist politics is that whilst embracing ecological modernisation and elements of progressive social movement agendas, contemporary democracies are failing to provide the 'level playing field' fundamental to developing environmental economies. It is only by beginning to unpick this paradox and the[se] tensions... that the transformation of eco-political (and wider social movement) agendas through selective accommodation within political systems and their recasting in terms of modernisation and progressive politics can be unpacked.[42]

In light of their work, I propose an alternative response to the situation that they diagnose. This book charts a course between the all-encompassing systems-functionalist objectivity that Blühdorn and Welsh offer, as well as the Scylla and Charybdis of Shellenberger and Nordhaus' realism and Swyngedouw's ultra-radicalism. In the next chapter, I develop an approach to the transformation of citizenship, the state and ideology over time that is critical yet pragmatically realist. The approach is designed to be *critical* insofar as it supports inquiry that aims to influence established patterns of domination in the direction of non-domination. It is designed to be *pragmatic* insofar as it recognizes the groundedness and boundedness of critique in what people do in society, rather than in abstract theoretical debates. As such, the approach is *realist* insofar as it is an exercise in crafting political judgement based in examination of what might be achievable, given current social conditions.

1.3 Citizenship, imperatives of state and ideological forms

Broadly Marshallian approaches generally discuss citizenship in terms of a historical narrative of change over time in the status *and* practices of individuals in relation to the state and other social institutions. Indeed, as an early theorist of the 'new' citizenship, Turner follows Marshall to define it 'as that set of practices (juridical, political, economic and cultural), which define a person as a competent member of society, and which as a consequence shape the flow of resources to persons and social groups'.[43] Although such views have been criticized for focusing on rights to the exclusion of dealing with citizenly duties, I accept Joaquín Valdivielso's argument from Marshall that the concept always bundles together questions of rights with those of duties, especially in relation to the category 'social' citizenship, which most interested Marshall.[44] Similarly, Engin Isin and Patricia Wood use Marshall's work to develop a more encompassing definition of citizenship as '*both* a set of practices (cultural, symbolic and economic) and a bundle of rights and duties (civil, political and social) that define an individual's membership in a polity'.[45] Broadly Marshallian theories of the greening or erosion of citizenship extend work by van Steenbergen, Turner, Isin and Wood, and Valdivielso, and include that done by Alex P. Latta, and Simon Susen, which I discuss in some detail.[46] By engaging with this research, this section introduces the theoretical framework that I will apply throughout the book to examine the historical establishment of civil, political, social and newer 'stakeholder' rights and duties against the backdrop of changing imperatives of state and a green ideological transformation from dualism to holism.

Moving to put aside an explicitly normative theoretical approach immediately raises three research problematics. The first problematic draws attention to clarifying how it is that increased awareness of society's damage to the ecosphere has, if at all, prompted a greening of citizenship as a status and practice in relation to the state. Valdivielso and Latta, in particular, share this view of citizenship as not merely being about 'its existing (or theoretically proposed) formal basis, but also ... the struggles that seek to reshape it'[47] in light of the ecological challenge. In this perspective, I seek to link new social movement and countercultural discourses with the key claims that I identify with the normative theories, and to describe how these may have become assimilated with the greening of citizenship. That is, I develop an approach to understanding how contemporary discourses are challenging the forms that discourses of justice and injustice take in the context of changing social conditions. The second problematic arises insofar as any observed transformations of citizenship might in turn be seen to affect a transformation in priorities, or what political theory regards as functional 'imperatives' of state. As such, I accept as given Turner's assertion that 'any theory of citizenship must also produce a theory of the state'[48] and set out an approach to this task. The third problematic is anchored by the question of what kinds of discourses and representations, shared or contested, might appear to motivate or undermine political support for the 'greening' of citizenship as a legitimate social goal. That is, it raises the problem of interpreting the ideological dimensions of the greening of citizenship *and* the state.

In order to address the first problematic, I follow closely the theoretical approach to citizenship that Turner and Susen prescribe. This means adapting a broadly Marshallian approach to understanding the development of citizenship rights and duties within the state over time, against the backdrop of a state-encompassing political culture. This approach involves qualitative interpretation of the *conditions*, *types*, *contents* and *arrangements* of citizenship[49] in terms that describe the establishment and transformation of different citizenship forms against a historical narrative of state formation. Specifically, accounting for the *conditions* of citizenship requires a focus on identifying what is spatiotemporally specific about the state formations that enframe novel forms of citizenship. Directing focus towards identifying the *types* of citizenship that manifest in a given space and time implies interpretation of the specific forms around which citizenly participation in social and political life is organized. A focus on the *contents* of citizenship concerns interpreting both the entitlements (rights, freedoms) and the obligations (duties, responsibilities) that arise from an individual's membership in a polity. A concentration on understanding the legitimacy of the *arrangements* of citizenship draws attention to the institutionalized ways in which the benefits and burdens of citizenship are distributed across different sectors of society.[50] The forms of citizenship that I describe using this approach relate to the mercantile state and bourgeois citizenship that arose in the 18th and 19th centuries, to the industrial state and social citizenship that emerged in the late 19th century and into the mid-20th century and to the emergence, since the late 20th century, of the postindustrial ecomodernizing state and what I will define as stakeholder citizenship.

In light of Turner's point that any theory of citizenship demands a concomitant theory of the state, the second problematic raises a need to interpret the impacts upon the state of what citizens may assert as rights and duties or, indeed, demand as institutional arrangements, as well as how the state and other institutions may respond to citizens as social actors. In this sense, the transformations of citizenship need to be understood in the context of and, in part, as affecting historical transformations of the state. In other words, a two-dimensional perspective upon citizenship and the state is required. What can appear from within one perspective as claims by citizens that injustice is present can seem from within an alternative perspective as incursions upon what are most often called prevailing 'imperatives of state'. Discussed in more detail in Chapter 2, 'imperatives of state' arise from the institutionalizing of the state around a particular 'set of individuals and organizations [that are] legally authorized to make binding decisions for a society'.[51] What is distinctive about 'the state' as a form of social organization is its historical establishment as that set or group of individuals and organizations within a society who actively assume authority that is exercised under constraint, whether that authority is delegated passively in a dictatorial manner or delegated actively by citizens in a democratic manner. The state is geared to the defence of functional imperatives that arise from the presence of historically entrenched structural constraints. Such functional imperatives are requirements that a state works to keep domestic peace, respond to external threats, prevent capital flight and raise revenues and, importantly, promote capitalistic economic growth and provide for democratic legitimation.[52]

However, the state, if it is to maintain stability in a society of citizens, must respond not only to functional imperatives that arise from the presence of historically entrenched structural constraints but also to some notion of justice. For Robyn Eckersley, the problem is that the basic functional constraints upon the state do not appear as 'objective constraints' to actors themselves. Rather, what one 'functionalist' kind of analysis regards as objective constraints need also to be considered in terms of actions 'filtered though the prism of different ideational frames by differently situated social actors within the state and civil society'.[53] Eckersley thus emphasizes the necessity and importance of a two-dimensional approach, insofar as focusing on a single dimension makes visible only a single aspect of the political narrative of change over time. Motivated by the presence of injustice or by changing circumstance, citizens can use the content of existing rights and responsibilities to *assert* their claims for novel types of social participation, rights and duties or institutional arrangements. Institutions can be seen to respond or not to these assertions, often in ways that bring about a transformation of the types, contents and institutional arrangements of citizenship themselves. In light of Eckersley's methodological point, I therefore add to Turner's and Susen's analytic categories of *conditions, type, content* and *arrangements* a focus upon citizenly *assertions* that injustice is present and institutional *responses* to them. Adding these categories allows inquiry to address the second problem that I identify by supporting reflection upon what Eckersley sees as constitutive acts – that is, assertions by citizens that injustice is present – and upon responses to them by the state, as that institution holding the legitimate authority to organize the rules for life held in common within a society.

Such a two-dimensional perspective is of particular importance when considering, as do van Steenbergen, Turner, Susen and others, the uneven influence upon citizenship and the state of the new social and broader countercultural movements that became established across the postindustrial societies in the 1970s and 1980s. Contemporary debate over the contours described by citizenship within the state have indeed become fraught in relation to this issue, especially on questions of how 'new' assertions of citizenly rights and responsibilities might have eroded state-based conceptions of citizenship as an empirical phenomenon *or* meaningful analytic category. One view in such debates is that citizenship, cast in terms of state-based civil, political and social rights and responsibilities, has in the late 20th century and into the 21st century come to lack practical or ideological relevance as a result of an economic prosperity-driven 'post-materialism',[54] which sees citizens abandoning concerns with obtaining riches and 'more stuff' and embracing intangible goods, such as free time, leisure activities, religious or quasi-religious spiritualism and New Age-type environmental values.

While I do not agree with this, the 'post-materialist thesis', I here put it aside until later and turn instead to Schlosberg's argument that, regardless of motivating factors, the entry into political debate of the new social movements since the 1970s and 1980s both broadens and deepens prevailing ideas of justice as something that citizens desire *and* the state as the imperative-bound container of the political sphere. In light of Schlosberg's work, it becomes important to discuss citizenship not only in terms of a distinct political sphere that supports debate over and upholds a definable and legitimate ideal of justice but also as something that emerges from within a broader cultural frame of reference. This is a matter of particular importance in relation to the third research problematic that I identify. Schlosberg's argument that the new social movements contributed to expanding what counts as injustice in society is helpful as it highlights the question of political legitimacy while tying that question to interpretation of how prevailing cultural forms may influence political institutions. This is because the new social and, importantly, broader, more diffuse countercultural movements have been central to shifting the focus of debates over justice in the context of unsustainable postindustrial development, *beyond* one-dimensional concerns with preserving states' rights or justifying 'pristine' nature[55] and towards the kinds of assertion that van Steenbergen and Turner identify. Hence, what is required is an approach that recognizes the terms on which new social and countercultural movements, in asserting claims for quality-of-life or security from risk, for example, contribute to re-defining the function of the state and justice. These movements emerge from within and appeal to a historically distinctive cultural ideological frame.

In Susen's terms, 'in order for the emancipatory potentials of a diverse and polycentric civil society [that has been established since the 1960s and 1970s] to have a tangible impact on the course of history, the discursive pluralism of new forms of collective mobilization needs to be [understood as having been] translated into the institutional pluralism of new forms of political organization'.[56] This characterization of the problem points back to Hayward's demand for a conceptual distinction between ethico-moral community and obligation, on the one hand, and political community and obligation, on the other hand. Insofar as new social and countercultural assertions that a situation is unjust emerged, they do not automatically constitute new forms of political community or obligations. New social and countercultural movements may sustain and support a civil sphere that has the potential to but does not necessarily shape political institutions. That is, the assertions of a movement are recognizably formalized only in the institutional arrangements of citizenship, just as for Marshall the rule of law before which all citizens are equal was established around assertions of civil citizenship, the principle of parliamentary representative government around assertions of political citizenship, and the welfare state institutions that redistribute material wealth around assertions of social citizenship rights and duties. When new institutional arrangements arise, such claims become visible as part of the political structure of the state, even if the state is, relatively speaking, weakened. Such caution creates a requirement for interpreting the *ideological* sources of citizenly demands, such as those emanating from within the new social and countercultural movements of the 1970s and 1980s. As Susen argues, to 'differentiate citizenship in such a way that literally any kind of social group can claim to institutionalize their collective necessities would mean to convert citizenship into a mere identity game on a higher level'.[57] That is, examining the transformations of citizenship over time necessitates a focus on the state as the primary institutional focus for interpreting the impacts of citizenly assertions that injustice is present. At the same time, recognizing that certain questions are forced onto political institutions necessitates an approach that can deal with the ideological dimensions of the citizenship issue in the context of the social whole.

This said, Latta also raises the possibility that something else is at work in the relationship between the greening of citizenship and the state. For him, 'citizenship does indeed, as Hayward suggests, have a fundamental link with the notion of a polity if we interpret this term loosely to indicate a collective that is defined in terms of a set of challenges held in common, over which shared rules of engagement prevail'.[58] However, Latta seems to weaken the concept of 'polity' to the extent that it incorporates or encompasses 'the physical, emotional and spiritual investments that people make in the ecological spaces where they dwell and work'.[59] In contrast, I regard such things as the stuff of the cultural, not the political, sphere. The point is that it is not so much the fact that 'shared rules prevail' but that some kind of community or institution is legitimately charged with organizing them which I take to be the defining feature of the political. With this distinction in mind, it is possible to view the achievements of the new social and especially the environmental justice and ecologism movements – which, for Schlosberg, have done so much to encompass the language and expand the discourse of justice[60] beyond hitherto prevailing definitions – as speaking from within a cultural ideological frame that has become attuned to and, in many cases, concerned with the impacts of environmental injustice or 'risks'. In this view, the citizenly actions of environmental justice and ecologism can be seen as manifestations of political community or obligations in and of themselves only when analysis distinguishes between the cultural and political domains of social reproduction. That is, recognizing the political impacts of citizenly actions requires a focus on assertions that injustice is present and institutional responses to these, and assessment in terms of the types, contents and arrangements of citizenship that become established in and around a shared cultural ideology.

Regarding them as analytic categories, I take *culture* to be the practices, discourses and material objects that express commonalities and differences, continuities and discontinuities of meaning over time within a society. As such, I regard *politics* as the social relations associated with legitimating the authority to organize the rules for life held in common in the context of a particular cultural frame. Culture provides the meaningful context within which 'society' and, so, politics, as well as the economy and ecological relationship with the ecosphere, are constituted. Hence, I seek to understand the political community and obligations that are constituted by the intertwining of citizenship with the state against the backdrop of a cultural grammar that facilitates social integration. I see citizenship and the state as expressing an encompassing cultural ideology that 'does not preclude contradiction and conflict'[61] but, in fact, encompasses or enframes political debate and difference at the necessarily abstract level of what political discourses have in common. Culture in this sense provides what Raymond Williams famously described as a shared structure of feeling.[62]

This distinction is central to the critical pragmatic political and social theory that has been developed by Boltanski, Chiapello, Laurent Thévenot and others, through engagement with work by anthropologist Louis Dumont and philosopher Paul Ricouer.[63] Their approach recognizes a need to distinguish between the *cultural* ideological 'whole set of representations that [can be identified as] facilitat[ing] social integration and identity preservation' in a given society, on the one hand, and the *political* ideological unmasking, distortion and dissimulation of social relations and conditions within a society, on the other hand.[64]

In this perspective, assertions that injustice is present and institutional responses to these, the contents of citizenly rights and duties and institutional arrangements, as well as the types of social and political participation of citizenship, arise within the framing context of a prevailing cultural ideology. Cultural ideology thus indirectly helps to define and legitimate the state as the authority that is charged with organizing the rules for life held in common in a particular society. Put differently, for assertions that injustice is present to effect the ordering of social reproduction, they must make political *and* cultural sense and, of course, attract the attention of the state, which may respond by establishing new institutional arrangements, which in turn help to legitimate it in the eyes of citizens. A corollary of this situation is that political claims arise from within a cultural milieu that motivates actors to act.

Differentiating between cultural and political ideologies in such terms follows Dumont's seminal work on ideological formation. Discussed in more detail in Chapter 2, Dumont's studies of ideological formation sought to understand the distinctiveness of Western modernization in comparative perspective, against the development of ideology in another major civilization, that of Hindu South Asia. In this respect, Dumont saw himself as extending beyond economic history Karl Polanyi's argument concerning the fundamental difference that arose between modern and non-modern societies as a consequence of what the latter describes as *The Great Transformation*. The present book critically engages the broad epistemological framework developed and expanded from Dumont's work in particular by Boltanski, Thévenot, Chiapello and the other French critical pragmatic political sociologists. The key idea in Dumont's approach to studying ideological formation is that of *l'englobement du contraire*: the encompassment by the whole of its contrary, which requires seeing the particular in terms of its relationship within the whole, in a way that defines the qualities of both. The aspect of Dumont's work that is most relevant to my argument is his conceptualization and detailed articulation of the cultural ideological sources of political problems, which in Western contexts he associates with *modern artificialism* or *dualism*: the radical separation of the Self from the World and Others that arose as a consequence of the secularizing of the Judaeo-Christian belief in the embodied and, so, uniquely *individual* Soul in conditions of modernization. The importance that Dumont gave to this concept represented his deeply held concern that, for more than three centuries, 'Western man [*sic*] has wanted more and more to be Prometheus, "lord and master of nature"'[65] but that, by the 1970s, 'protest against the disruption of natural equilibriums [had] for the first time in public opinion [arisen and] placed a check on modern artificialism.'[66] Dumont's argument recognizing the radicalness of modern artificialism as a distinctive cultural ideology, established as a condition of modernization in the West, provides the backdrop for this book. It supports my effort to better understand how cultural ideological shifts sometimes boil over into political tensions, and how the actions that actors take in particular social situations can, over time, transform citizenship and the state.

Seen in these terms, ideology provides a framework for debate over the legitimacy of the state as the primary political community while also informing what it means, in terms of sharing a culture, for citizens to contribute to society in general. I develop this perspective because, as Isin and Wood argue, it is necessary to view citizenship not merely as a legal or political status or as the basis for political engagement but as a 'way of life' or framework for cultural, symbolic and economic practice.[67] Such an approach also complements what Eckersley defines as the requirement that analysis take into account the relationship between functional and constitutive dimensions of politics, as well as including a means for interpreting prevailing and emergent 'ideational frames'.[68] This means that cultural ideological and political ideological differences, citizenship and the state itself, need to be considered within the frame of what Thévenot regards as a 'duly qualified reality'.[69] The key premise here is that actors, acting from within a particular cultural ideological frame, face political requirements that they give reasons for their actions, which are set against the backdrop of historically established social conditions, relations and reproductive practices. The practices and discourses that political actors put forward in social situations represent the reasons or 'justifications' that they offer for their actions. Such justifications draw on a cultural ideology that is forged by people over time in the material world, thus 'qualifying' reality. As Boltanski, Thévenot, Chiapello and others demonstrate, in order to be accepted by others as valid reasons for action, justifications must appeal to the general interest or common good: *justification* is the culturally sanctioned objective of debate in the political sphere of the state.[70]

In Chapter 2, I elaborate this approach, arguing that the state is geared not only to defending certain functional imperatives that arise from the presence of historically entrenched structural constraints but also to preserving some observable culturally legitimate representations around which political debates over justice are contested and defined. That is, political actors can be regarded as being bound to provide justifications for acting, such that by asserting that injustice is present they also potentially define the cultural ideological 'grammar' that provides the basis for describing or representing the general interest or common good. For Boltanski, Thévenot, Chiapello and others, the need for justification accompanies the emergence of modernization as a social-historical process and is central to the reproduction of state-based societies that privilege formal equality amongst autonomous individual citizens. For them, the need for justification is therefore a condition of secular pluralism, insofar as a justification for action tends to follow one of several distinctive orders of worth, orders of justification or *models of justice*.[71]

For this group of theorists, each model of justice offers an analytic category that describes the reasons that actors give for their actions, implicitly or explicitly, when addressing social issues. For example, in debates between 'humanists' and 'theocrats' in post-revolutionary France, or in the trade-offs between profit-sensitive real estate developers and environmentalists sensitive to the intrinsic value of 'untouched' wilderness in the contemporary United States. The key point made by Boltanski and Thévenot is that even though at any one time a vast multitude of *political* ideological models of justice may obtain, at any particular historical moment it is likely that a single cultural model of justice predominates because it is most effective in relation to and, as such, in part qualifies social reality. A prevailing cultural ideology and social conditions within a state, taken together, *qualify* the reality against which political discourses of self-interest or altruism or, indeed, efficiency, solidarity, trust, creativity, charisma, flexibility or unity with nature, for example, can be evaluated as contributions to a cultural grammar that frames the contents of justice and injustice. Such categories are designed to account for the demands that politics places upon diverse actors, against the backdrop of a prevailing cultural ideology and the institutionalized political community of the state. The categories are 'models of justice' in that they offer people a means of understanding and judging the relative validity of different facts and ideas, the value of different individuals and groups, and their own value and potential role in political debates, as they constitute 'reality'. Being part of a shared culture, assertions that injustice is present contain and express principles of equivalence that have a claim to validity in society at a given point in time and are oriented, in some sense by their very structure, towards universal validity. The critical pragmatic approach does not attempt to interpret the psychological motivations of actors, nor does the approach require axiomatic assumptions about self-interest or altruism. Justification is merely discussed in terms of a public representation that is notable because it attempts to move beyond 'a particular or personal viewpoint toward proving that [a] statement is generalizable and relevant for a common good, showing why or how this general claim is legitimate'.[72] The concept of justification offers a means by which discourses might be open to interpretation as contributions to social life, whether such discourses might be reactionary, seeking to justify existing structures of privilege, or progressive, seeking to justify the expansion of the benefits of social existence to all on an equal basis.

This said, there remains some debate within the critical pragmatic camp on exactly what constitutes *the* prevailing model of justice in contemporary society. I am unaware of mutual agreement between Boltanski and Chiapello, who advocate the position that a unique 'connexionist' model of justice prevails in the contemporary West, and Thévenot, with Michele Lamont, Claudette Lafaye and Michael Moody, who seem to advocate the position that the most recent shift that exerts an influence upon political debate, in the United States and France at least, is that of 'greenness'. In light of this as-yet-unresolved tension in critical pragmatic theory, which I discus in detail at a number of points in subsequent chapters, I step back from their multiple models of justice approach in this book. Rather, I favour a broadly pragmatic framework that sees actors as the inhabitants of social networks, in and around which opinions on the justice or injustice of an action tend to aggregate. Or, less abstractly and concerned primarily with developments in recent decades, Pierre Lascoumes' argument that the impacts upon culture, politics, the economy and society in general of the environmental movement and the normalizing within contemporary culture of environmental awareness have created a situation in which 'green' concerns are exemplified by a range of institutionalized, quasi-institutionalized and ephemeral 'actors' who, holding conflicting interests and strategies, bear back upon all social institutions and, in particular, the state as environmental issues are 'integrated' into general policymaking. For Lascoumes, the postindustrial state in particular plays an important role in normalizing 'greenness' by managerializing environmental problems, while simultaneously this managerialism, by promoting measurement, predictability and control of such problems, configures or constitutes nature as a visible product of social relations, rather than as something dominated by them.[73]

In light of Lascoumes' work, which in a certain sense resembles Schlosberg's analysis of the 'expansion' of the concept of environmental justice by citizens acting in new social movements, and in contrast to Boltanski and Chiapello and, to a lesser extent, Thévenot, Lafaye and Moody, I develop a dialectical interpretation of the shifting influence of holistic and dualistic cultural ideologies upon political ideological formation. Without further discussion of this central motif here, my point is that rather than focusing on the dynamics of debate between and across an array of different political models of justice, I focus more closely on how changing cultural ideology over time can create demands that progressive and reactionary political ideologies define and redefine themselves, thus 'qualifying the reality' within which forms of citizenship and imperatives of state emerge.

This said, I view the political discourses that can be said to arise within a cultural ideological frame – such as over the rights and duties of citizenship or the types of social and political participation that institutional arrangements should support – as revealing of what Boltanski and Thévenot regard as *compromises*. Compromises have two qualities. First, they arise as practices and discourses justifiable by reference to one set of political representations overlap and are made compatible with those of another, against the backdrop of a cultural ideologically qualified reality. Second, while there exist significant tensions between the different courses of action that a compromise situation encompasses, an intentional proclivity towards the universality that a shared culture affords is maintained, without addressing political tensions directly: 'without [attempting] to clarify the principle upon which ... agreement is grounded'.[74] In situations characterized by compromise, attempts to reinforce one or another political ideological position or to map out an alternative model are marginalized from politics, but may remain salient in cultural terms. A compromise arises when different conceptions of how best to organize the rules for achieving the common good gain legitimacy within a prevailing cultural ideological frame without necessarily being identifiable as coalescing in a way that changes the prevailing cultural ideological grammar.

As I propose in Chapter 3, observing the conditions, assertions, responses, contents, types and arrangements of citizenship in such terms highlights the working-out of political compromises against the backdrop of a prevailing cultural ideology and amidst the inertia of functioning state imperatives. As such, the key to understanding transformations of citizenship and the state over time from within this perspective lies in recognizing that such compromises are inherently unstable. To evoke another well-known example, the institutional arrangements of what Marshall saw as social citizenship, when regarded as providing contingent resolution to political tensions against the backdrop of the industrial struggles of the mid-20th century, can reveal the persistence of cultural ideological dualism. The tensions between reactionary and progressive political ideological positions were left latent or resolved around the question of redistributing the spoils of industrialization, a formal process through which the state was able to shore up its legitimacy by institutionalizing 'class conflict' in the arrangements of the welfare state. During the Cold War, the rights and duties of social citizenship provided a formal framework for debate between competing progressive and reactionary political movements. The advocates of progressive and reactionary positions settled upon a welfare state 'compromise'. Throughout the Cold War era, social democrats supported greater collective solidarity, while reactionaries supported self-interested atomism. Seen from the functional perspective of state theory, progressive and reactionary groups were engaged in efforts to align their interests with state imperatives, while the state sought to maintain legitimacy by providing a welfare state that, alongside more radical or extreme visions of the common good, helped to shape cultural expectations. From within a constitutive perspective, a dualist cultural ideology that gave meaning to issues of redistribution supported a range of reactionary and progressive assertions that injustice was present, not all of which could gain purchase in the political sphere but many of which did, shaping the contents of social citizenship for several decades. What facilitated the establishment of this compromise was the crystallization of politics around shared, in some cases explicit and in some cases implicit, acceptance of a dualist cultural ideology. It was an 'us versus them' culture of competing social classes – one good, the other bad; one on the side of progress, the other reactionary – which gave the politics of redistribution meaning. The theoretical point that I draw from this brief *vignette* is that there exist significant practical and discursive tensions between the different courses of action that a compromise encompasses, and that these can be observed and interpreted only against the backdrop of an account of cultural ideological representations.

Such an approach holds, at least, in the sense that cultural ideology and social conditions combine to define the parameters of a *test* of the justifiability of political assertions that injustice is present. This explicitly heuristic approach reveals the emergence of conditions for a test that establishes, by exclusion, what are more or less justifiable political claims. Returning to the example, it might be argued that in the Cold War, key cultural and political questions were satisfied through the establishment of institutional arrangements of social citizenship that formalized the redistribution of economic wealth. As such, there is an analytical need to identify what a *test* specifies; in this example, it is the possibility that questions of justice could be settled through the redistribution of wealth. A test in this sense provides the political means for ranking status in society and signals to inquiry the prevalence of a distinct set of cultural ideological representations. In contrast with non-modern societies, in which reality furnishes the unquestionable Truth against which human actions are judged, with the advent of modernity reality itself is open to definition and redefinition. In non-modern societies, an individual or group of individuals holds privileged, sanctified access to cosmological knowledge, and so power, concerning the divine origins of society and the rules by which it is to be reproduced. In modernity, the presence of a secular and, so, mutable cultural ideological grammar means that the tests in relation to which political differences are settled require the presence of a legitimate arbiter. Put briefly, in abstract terms, a test requires a politically sanctioned and culturally legitimate adjudicator. The arbiter of the test is, in principle, citizens acting collectively. However, in practice, it is the state, which acts to manage the compromise on the social bond between individual autonomy and collective obligation by settling political differences against the backdrop of a shared cultural ideology. This is not to argue that the state is neutral, but only that it may be as good as its citizens' capacity to steer its objectives.

The take-away point is that tensions between different political ideological positions can, when the stakes are high, coalesce in a reality-changing or re-defining moment, with the result being the establishment of a new test.[75] Where a compromise breaks down – and the result is neither violence nor, indeed, the collapse of the state – recognizing the establishment of a new test effectively reveals the qualification of a new cultural ideological grammar and a new 'reality' against which new forms of political representation can emerge. The qualification of a new reality brings about a situation in which assertions that injustice is present come to be cast in a new grammar or vocabulary: that is, using appeals to novel or hitherto marginalized cultural ideological representations. In Chapter 3, I examine in admittedly approximate, narrative terms the destabilizing of cultural ideological dualism and with it the welfare state and social citizenship with the establishment of the postindustrial state. I find that, from the 1980s onwards, the compromise that had been settled around the test of wealth became increasingly unstable. The greening of citizenship in this sense does not alter what is essential about progressive and reactionary political ideological positions. However, it does serve to delegitimize the test of wealth, while evoking a new test of wellness that, potentially, expands possibilities for asserting that injustice is present. That is, the inevitable destabilizing of a compromise over time provides an insight into discourses that are re-defining citizenship, the state and ideology. The chapters in Part II combine observations on the conditions, assertions, responses, contents, types and arrangements of citizenship with a focus upon cultural ideological representations, and the transition between compromise and test, placing a particular emphasis on the implications for thinking about differences between reactionary and progressive political ideological positions (see Table 1.2).

Table 1.2 Thematic framework for analysis

Different citizenship forms emerge out of spatiotemporally specific social *conditions*. The conditions of citizenship can be identified with the social forces that are produced and sustained by virtue of state power.
Assertions of citizenship constitute the claims that characterize struggles to define the status of individuals in relation to the state and other social institutions.
Institutional *responses* to citizen assertions that injustice is present hedge against or further the allocation of entitlements and obligations.
Types of citizenship refer to the specific forms in which social and political participation is organized.
The *contents* of citizenship concern both the entitlements and the obligations which arise from an individual's membership in a polity, and represent the exact nature of the rights and duties which define citizenship.
The *arrangements* of citizenship constitute the institutionalized ways in which the political benefits and obligations of citizenship are distributed within society.
A *test* provides the criteria by which status is judged in relation to the social whole. A test is brokered by the state, acting as the legitimate authority charged with the role of arbiter of the compromise on the social bond.[a]

Sources: [a] S. Susen, 'The Transformation of Citizenship in Complex Societies', *Classical Sociology* 10, no. 3 (2010): 259–85; B.S. Turner, 'Outline of a Theory of Citizenship', in *Citizenship: Critical Concepts*, ed. B.S. Turner and P. Hamilton (London: Routledge, 1994 [1990]), 199–226; L. Boltanski and L. Thévenot, *On Justification: Economies of Worth*, ed. P. DiMaggio et al., trans. C. Porter, *Princeton Studies in Cultural Sociology* (Princeton: Princeton University Press, 2006 [1991]); M. Gauchet, B. Renouvin, and S. Rothnie, 'Democracy and the Human Sciences: An Interview with Marcel Gauchet', *Thesis Eleven* 26 (1990 [1988]): 147.

1.4 Crafting political judgement

What I suggest is taking place in the early 21st century is a cultural and political transformation that the normative theories of green citizenship exploit only partly. New social movement demands for environmental friendliness and ecosystem balance, cast alongside less coherent countercultural demands for authenticity and creative expression, have destabilized dualism as a cultural ideology. Not merely the possibility of compromise around what I will describe as the test of wealth and the institutional arrangements of social citizenship and the welfare state have been dissolved. In Chapter 4, I describe how, through the establishing of new types of social participation, and rights and duties and institutional arrangements of *stakeholder* citizenship in the 1990s and 2000s, the practical implications of this cultural ideological turn from dualism to holism become visible. I propose the view that stakeholder citizenship has become established as a partial consequence of institutional responses to new social movement and countercultural assertions that injustice is present, especially through political party embracing of local communities, bonds of trust and 'workfare not welfare'.

In particular, I examine how the embrace by the postindustrial state of neoliberal ecomodernization policies in the face of subpolitical concerns with wellbeing and risk issues, amongst other things, is contributing to the qualification of a new test of wellness that partly and problematically assimilates the five key claims of normative theories of green citizenship. That is, I describe the shift from a relatively unambiguous dualist cultural ideology, one that gave meaning to economic redistribution as a justice issue, to a relatively ambiguous holistic cultural ideology. Political debates in the 2000s tend to coalesce on questions that extend from the injustices of economic maldistribution to the misrecognition or invalidation of identity, misrepresentation or lack of participation and, ultimately, encompass human flourishing or capabilities as well as societal capacities to contribute to sustainable development. Following Fraser and Schlosberg, amongst others, I find that this expansion and reframing of justice claims may incorporate many possibilities for achieving the green ends of strong sustainable development. However, it is not necessarily the case that the kinds of ends that greens advocate take precedence within a holistic cultural ideological frame. As well as strong sustainable development, a holistic ideological grammar can also frame politics in terms of access to astroturfing green consumerism or greenwashing 'market globalism'[76] that displays few or even negative prospects for achieving sustainable development.

This situation has an acute effect on the ways that politically progressive and reactionary positions are enunciated in the postindustrial ecomodernizing state. In contrast to arguments that the new citizenship dissolves differences between the two value orientations, I find that the incorporating of new social and countercultural movement claims into the contents of contemporary citizenship has simplified reactionary appeals to preserve existing structures of privilege, which are now made directly on the basis of subjective freedom of choice to consume resources and social policy that will further 'unburden' individuals and communities. Meanwhile, I find that progressive aims of universalizing the possibility of freedom by equivalizing economic redistribution are complicated, and require considering a more 'whole' conception of justice.

In Chapter 5, I draw into question what it means to assert that injustice is present against the backdrop of a holistic cultural ideology and postindustrial conditions. Moreover, I concentrate upon the forms that injustice takes in relation to the rights and duties and institutional arrangements of stakeholder citizenship. I develop the view that, in the society of stakeholder citizens, if the state is to continue to legitimately organize the rules for life held in common, it is bound to increasingly take on the task of supporting individual and community self-responsibility in relation to questions of wellbeing and its correlates, independence and security from risk. I argue that, unlike the dualist concept of *exploitation* that was central to the politics of distribution and the institutional arrangements of social citizenship, injustice in postindustrial conditions is represented in holistic terms and, as such, has no visible class of *perpetrators* and is *diffuse* in terms of responsibility for it. No single class or group can be held responsible for low levels of wellbeing or high levels of risk, dependency or related issues such as low social mobility, participation or inclusion or, indeed, unsustainability. The key exception, as Honneth and others such as Zygmunt Bauman and Richard Sennett note,[77] is that individuals themselves are held responsible for failing to take advantage of the opportunities that states and markets cooperate to offer. Indeed, I argue that issues closely relating to the types of social and political participation that are central to contemporary stakeholder citizenship, such as concerns with liveability, are problematic insofar as those who can be held directly responsible for them are those who make bad life choices, are unwell and consume the wrong products. Similarly, if an individual does not have an income in the contemporary state, it is increasingly seen to be because that individual is not self-responsible and attentive to all of the dimensions of wellbeing, such as being independent, creative, innovative, entrepreneurial, motivated and pro-actively engaged in the pursuit of personal excellence. To illustrate my point here, such a shift sustains the narrative in television series such as *The Wire* and *The Sopranos,* in which criminal activities provide the allegory for individual attempts to flourish in a postindustrial 'world without work', where how one obtains an income is seen both subjectively and objectively as a lifestyle choice. Conversely, personal choices to consume with green or social awareness – the *Stuff White People Like* and do – tend to be over-valued as contributions to political community or as fulfilment of political obligations. Hence, Chapter 6 examines how some well-known movements – such as advocates for environmental justice, a financial transactions taxation, guaranteed basic income, fair-trade agreements between nations and social indicators of sustainable development – might be thought of as challenging the forms that injustice takes *after* dualism.

Therefore, while I see normative theories of green citizenship as highly important and indeed indispensible tools for debating and reflecting upon the kinds of means and ends that greens should support, this book is designed to complement such efforts by evaluating the possibilities and bottlenecks that existing institutional arrangements and ideological forms present before such a project. I see green citizenship theory and practice in the same terms as Zev Trachtenberg, who regards it 'a descriptive enterprise; it serves to help people better understand what they already know'.[78] Such a framing of the problem highlights the importance of debating and working to implement green political objectives amidst a postindustrial state that is bound to arbitrate the compromise on the social bond around a test of wellness. To paraphrase political philosopher Raymond Geuss, the aim of political analysis in modernity *might at best* be to explain what it is that moves people to act in a particular, historically framed social context, and to observe and critically interpret 'what then happens to them and to others as a consequence of how they act'.[79] For Geuss, political analysis should locate social practices and discourses historically and ideologically, and in doing so draw attention to actions and the contexts of action, rather than to subjective opinions or beliefs. Or, again paraphrasing Geuss, political analysis in this view might aim to 'craft' an adequate critical description of what is, and use this as the basis for a realistic and critical account of what might be possible.[80] Informed by these reflections on the task of political analysis, this book is an exercise that aims at contributing to the crafting of political judgement. Unfortunately, it appears that little can be done to steer Western societies onto a sustainable development trajectory, at the level of citizenship, the state or globally. Worse still, even if such a shift were to take place, the problems that are posed by the situation in the non-Western world would remain unaddressed. I believe that neither of these observations provides sufficient reason not to act. In this view, beliefs that another world is possible might be well served by placing emphasis, not on describing that other world, but on exploring what might be possible from within this one.

2
Modern Artificialism: An Alternative Perspective on Nature/Culture Dualism

2.1 Modernization: from holism to dualism in Dumont

Drawing on secondary and theoretical literature, discussion in this chapter builds upon Dumont's concept of 'modern artificialism' to focus upon the links between citizenship, the state and what is most often discussed in green political theory as nature/culture dualism. In particular, the discussion draws attention away from the dualistic split between nature and culture. This approach serves to highlight links between the historical emergence of a dualist or artificialist modern culture, on the one hand, and a counter-mandatory holist cultural modernism, on the other hand. Using Dumont's work in this way also focuses inquiry on the tension between reactionary and political ideological positions amidst the establishment of modern bourgeois citizenship and the mercantile state.

For anthropologist Chandra Mukerji, it was 'the plethora of objects arriving or produced in the West' from the 16th and 17th centuries onwards that was central to the process of modernization that, once established in Europe, was exported to the New World. Mukerji paraphrases Polanyi and Marshall Sahlins to describe as 'cultural materialism' the defining feature of an emergent 'social system in which material interests are not made subservient to other social goals'.[1] As hitherto localized and relatively self-sufficient communities expanded to encompass more space, modernization as a social-historical process inveigled more people, goods and services, in greater concentrations, than had agrarian societies. Simultaneously, many aspects of social reproduction were mechanized, while common land was enclosed as private property and developed for intensive cropping and industrial use. Towns and cities grew exponentially as hitherto unimportant organizational domains became more important to social reproduction. Notably, capitalistic markets and mediatized modes of communication and administration made relatively de-personalized forms of authority and abstract criteria for determining value central to social life; time and space were liberated from the sacred and eternal domain as value-free clock-time and cadastral space, while the ecosphere included less that was perceived as being immeasurable and unquantifiable.[2]

That is, the arrival of such a plethora of objects brought relatively new ways of understanding and perceiving the world, which characterize the modernization of Western social relations:

> Where the sheer number of objects available to people and the increasing manufacture of goods for practical use – such as clocks, books, maps, and guns – also began to affect practical activities ... [people] became less dependent either on other persons or on nature [or cosmology]. Once [material culture] is produced, it is part of the world in which people must function (at least, until it is destroyed or replaced with other goods) and to which they must adapt their behaviour ... [T]he expansion of trade and increased appearance of objects on the market were the occasion for the establishment of new and elaborate systems of thought, ones that advocated careful measurement and study of relationships among variables, conceived of as material forces.[3]

These changes affected a shift away from a traditionally ordered holistic relationship between society and the ecosphere, between persons and the natural world inhabited by them, and among persons. Cosmologically given Truth gave way to what Max Weber famously defined as an arbitrary and contingent, yet potentially measurable and, so, knowable, universe of material forces.[4] Mukerji's suggestion that persons 'could turn more frequently to things' helps to explain how a particular form of life, one that involved historically peculiar orders of skills, dispositions and understandings, became central to modernization in the West. As persons 'become less immediately dependent on others or material environments', they came to reckon temporality in precise increments; to see the body as the receptacle of labour-power, an objectively quantifiable thing; to recognize space as empty and appropriable property or territory. Meanwhile, information about the material world became relatively more specific, as well as intensive and extensive.

All-encompassing categorical aspects of lived experience – such as space, time and knowledge-belief, as well as embodiment itself[5] – came to be seen as the self-evidently immutable properties of an objectified natural world. Expanding capitalistic markets and urbanization increasingly called upon persons to manage future uncertainties as problems in the present. Modernization required the autonomous application of personal judgements based on worldly observation, rather than transcendent inspiration: selfhood came to be experienced as an apparently fully autonomous subjectivity. The secular materialism that arose impelled persons to exercise individual autonomy, to acquire skills and dispositions, to assume the knowledges necessary for engaging in a social universe that was experienced as something external to the Self. This situation differed from the non-modern experience, wherein the subjective self was enmeshed within a cosmologically framed eternity.

As the techniques and knowledges associated with the uses of material goods became more important within society, this relatively direct union of the embodied soul with the transcendental divine gave rise to what Dumont describes as the 'peculiar ideology of Western individualism'.[6] That is, the plethora of goods that is described by Mukerji arrived into a Judaeo-Christian Western culture, one that had historically placed great emphasis upon the self-salvation of the individual soul through worldly actions. Dumont sees the emergence of Western individualism as a condition of the transformation of this aspect of Occidental Judaeo-Christian culture. He describes a process of cultural-ideological syncretism that brought atomized subjects, charged with a duty to save their own souls through worldly actions, to the centre of an objective nature. Dumont's primary focus is upon how Western culture dealt with the ever-expanding scope of modernization, in particular with capitalistic markets and industrial production. Amongst other things regarded as important by him was that these made possible the wide-reaching dissemination of news-type information, which meant that persons could identify pan-European clerical corruption and venality with Papal Rome, and unjust social conditions with aristocratic control of land and people.

For him, the flourishing of Lutheran and Calvinist protest against the Roman Church altered how the Judaeo-Christianity that was central to Western cultural and political life was lived. Dumont reveals a situation in which persons, confronted with knowledge that Church corruption extended all the way to the top – and increasingly acting as 'free' labourers, artisans, entrepreneurs and intellectuals, as opposed to peasants, serfs, lords and clergy – began to entertain a relatively direct union with the divine. Individuals, acting autonomously as embodied individual souls, legitimized a private inner-worldliness that could oppose the public degradation by Rome of religious legitimacy and the depredation of 'natural order' by a supine aristocracy. In this atmosphere, dogma pertaining to religious observance and subsequently to feudal authority came to be seen as a constraint upon individual persons' 'peculiar subjection to God's will': the belief-system that once offered 'refuge from this imperfect world in another [transcendent] one' underwent a metamorphosis, which produced the 'Western ideology of individualism' and its cultural correlate 'modern artificialism'.[7]

Dumont describes how individuals came to inhabit what Weber describes as a subjectivist culture, a disenchanted world of potentially knowable dimensions that are seen to exist of themselves, devoid of the ethico-moral trappings that a divinely ordered cosmology entails.[8] Nature came to be subjectively experienced as something comprehensible on objective terms: 'artificially', as matter, forces and energy. What had hitherto been divine Truth in the holistic sense of an all-encompassing cosmology became 'concentrated in the individual's will', a dualistic metamorphosis that Dumont describes 'as the model of modern artificialism at large':

> [Modern artificialism is t]he systematic application to the things of this world of an extrinsic, imposed value. Not a value derived from our (i.e. humanity's) belonging in this world, such as its harmony or our harmony within it, but a value rooted in our heterogeneity in relation to it: the identification of our will with the will of God. The will applied to the world, the end sought after, the motive and inner spring of the will are extraneous; they are to say, the same thing, essentially outworldly [yet] concentrated in the individual's will.[9]

In short, modern artificialism is the radical dissimulation of Self from World and Others, the generalizing of which heralds 'the West's Promethean moment'.[10] Marcel Gauchet takes up Dumont's interpretation of the impacts of modernization, recognizing that '[t]he term "lord and master of nature" does not simply represent a certain level of scientific, technical and economic development but much rather expresses a symbolic organization of experience, where material abilities and intellectual attitudes cannot be separated'.[11] Modern artificialism is, in brief, the application to worldly affairs of personal will-objectives that are, in effect, 'the concentration within individuals' wills of objectively true reality'.[12] Moreover, and importantly, displaced are traditional ancestry-based social hierarchies that depend upon how a person *seems* in relation to past and tradition. In their stead arise the somewhat less definitive class divisions that depend upon what a person *is* in relation to the present.[13] In the context of a holistic cultural ideology, hierarchy establishes an actor's status within the cosmologically determined eternity that configures time, space and embodied presence unitarily. In the context of the modern dualist-artificial ideology that was established, hierarchical status gives way to formal equality among 'free' actors, who confront others and nature as social atoms, and constitute the elementary particles of society. With the advent of modern artificialism, subjectivity is established as an autonomous and self-contained agency that confronts an objective field of forces.

In the context of what Polanyi saw as the radical 'disembedding' of economic from cultural life, a radical transformation of the relationship between society and the ecosphere that encompassed it and of the relationship between people within society spread across Europe and the world. Modern dualism or artificialism emerged as the central principle of social reproduction, such that individuals were called upon to relate to each other and with the non-human world in instrumental terms and, as a consequence, increasingly came to identify their own wills as the self-sovereign expression of autonomy. A 'shift in the focus of domination' was established with the emergence of modern artificialism, such that the immutable pathways of hierarchical domination *within* an all-encompassing social bond were displaced by mutable patterns of domination *over* an objectified external natural world. In this view, as the non-modern cosmologically ordered relationship that links nature with the past is dissolved, modernization creates a 'loss of responsibility for social cohesiveness'.[14]

As individual autonomy emerged as the condition of social engagement, the generalizing of modern artificialism pitted atomized subjectivity against an alienated and externalized universe. Whereas domination had previously 'passed through the medium' of holistic, eternally maintained and hierarchical social relations, 'it was now [in conditions of modernization] presupposed on the isolation of the actor confronted with nature'.[15] Whereas in holistic societies, social relations were grounded by the timeless order of things, in the individualistic social formation that was established, social relations came to be anchored by domination over material things, such that 'alienated' relations arose amongst formally 'equal' individuals and nature itself. Whereas traditional societies had worked to maintain harmony within an eternal order of things, at the cost of permanently unequal hierarchy, modern society and the individuals constituting it work to perfect the world and the self on the basis of a generalized self-understanding of the naturalness of inter-individual equality, freedom and autonomy, regardless of whether such an 'ideology' is only formal and involves the destructive exploitation of others or the ecosphere.

More recent work by Biro helps to flesh out some of the implications of the impacts of modernization in slightly different terms. For Biro, in modernizing conditions the 'biological instinct for self-preservation',[16] hitherto constrained by an all-encompassing and explicit holism, is cut loose and individuals are effectively left to their own devices amidst a nature that emanates objective, positively knowable laws. In conditions of societal modernization, binding, traditional holistic norms – including explicit recognition of the 'normality' of social hierarchy and inequality – begin to manifest as limitations upon autonomous and self-orienting individuality. Dualist-artificialism displaces a binding totality that confirms subjectivity in relation to cosmologically ordained and unquestionable myth and tradition. In this argument, and following Biro, self-preservation is not necessarily identical with self-interest. Self-preservation implies the basic tendency for any biological organism to seek to go on in a given context. In traditional holistic societies, self-preservation takes the form of immersion or embeddedness in a collective way of life that is attuned to the maintenance of the cosmological order. In contrast, self-interest refers to the kinds of practice that self-preservation tends to involve in modernizing conditions. In conditions of modernization, a person, faced with adversity, is artificially set against an objectified universe of material forces, and must decide what to do *as if* for themselves, as must everyone else. This is what the left-Hegelian critical theory that is the topic of Biro's research, especially as expressed by Theodor Adorno and Max Horkheimer, views as the potential of modernity to facilitate a potential 'dialectic of enlightenment'.[17] In contrast to the critical theorists, this condition is also what the classical political economists would come to regard as the quintessential human condition: atomized individuals bound over to maximize personal utility amidst an indifferent nature. As such, it is possible to think about the transformation of Biro's biological instinct for self-preservation in two registers. Broad self-interest manifests in terms of what the Frankfurt School theorists regard as critical thinking towards progressive, collective 'enlightenment', while narrow self-interest manifests in terms of reactionary defensiveness of existing personal or group privileges.

Hence, Dumont's argument is that modernization ushers in a comprehensive cultural ideological turn 'away from holistic society and towards individualism, which emerges as the central principle of societal reproduction' and 'the highest social value'.[18] Or, for Alain Supiot, 'in most other civilizations people consider themselves to be part of a whole which both surrounds and goes beyond them'; in the West it is the autonomous individual that emerges as the 'elementary particle of society'.[19] With the advent of modernization as a social process, individuals in this sense engage in culture and politics in historically distinctive terms. That is, individuals are called upon to act *as if* autonomously set against an objective natural world and appear culturally and politically as if they were equals, peers engaged in an 'artificial' struggle with things, and others. For Gauchet, the transition from traditional holism to modern artificialism meant that what had once been sacred and external to the self became internalized, 'into everyday experience itself, into social ties, into relationships with nature, into the psychical constitution of human beings'.[20] Modern artificialism represents the collapse of the social organization of otherness in holistic terms and the internalization of otherness, such that social divisions hitherto organized in terms of hierarchy and dependence morph into divisions within the psyche and independent relations between formally free and equal, autonomous individuals. Failure and success are no longer consequences of choices made in relation to eternal Truth but consequences of subjective value appraisals made in relation to a given situation. Support for existing social structures no longer expresses support for an eternal, immutable Truth, but support for existing structures that privilege certain ways of doing things and, most often, certain structures of privilege. Similarly, rejection of existing social structure no longer implies a rejection of society itself, giving rise to banishment, but implies a desire to challenge the social structure in order to improve present conditions into the future.

This artificial separation of the self from others and nature accompanies the normalizing of a historically novel form of individual self-sovereignty and a novel institutionalization of the state. By marking the radical separation of the self from others and nature, artificialism brings individual self-sovereignty to the centre of social relations. This 'ontological' division of the universe across subjective and objective domains made individual self-sovereignty an indispensable quality of social existence, and presaged the establishment of 'citizenship' in the modern sense, as it is discussed by Marshall, Turner, Isin and Wood, Susen and the green citizenship theorists. This understanding of nature/culture dualism as modern artificialism partially reframes what Marshall sees as the condition in which citizenship that is premised upon 'the right to [secular] justice' became central to politics and the modern state.[21] A secular, alienated and artificialist culture that privileges individual ethico-moral choice and self-sovereignty sustains the political differences between reactionary and progressive political positions. The concept 'modern artificialism' as such helps to frame the Marshallian insight that 'the emergence of the modern citizen requires the constitution of an abstract political subject no longer formally confined by the particularities of birth, ethnicity or gender'[22] and bound to act *as if* autonomously set against an objective natural world while appearing each to another as – at least formally – a free cultural and political equal.

2.2 Modernization, citizenship and the state

In this perspective, the emergence of modern artificialism as a cultural ideology created the formal possibility that existing conditions could be confronted as unjust for earthly reasons. The establishment of an artificialist culture fostered a situation in which autonomous individuals increasingly came to assert claims that political authority legitimates itself before them as a 'public'. A key impact of the emergence of bourgeois citizenship in the mercantile state was that political actors, seeking power and influence, became bound to demonstrate within an emergent political 'public' sphere that *their* policies would support shared cultural ideals of freedom, equality and autonomy while also addressing key imperatives of state. The state in this view emerges not only as the authority legitimately charged with making binding decisions on behalf of citizens but also as the arbiter of a compromise on the social bond, between collective obligations to value individual freedom, equality and autonomy, on the one hand, and the demands of all individuals, bound to 'act out' this freedom, equality and autonomy, on the other hand. This is the central issue of concern in Jürgen Habermas' seminal work on the 'structural transformation of the public sphere' in the mercantile state. I now turn to situate Dumont's account of modern artificialist cultural ideology in relation to Habermas' thesis. This draws the concept of artificialism or dualism into Marshallian arguments that claims for citizenship sought to define justice in terms of *civil* demands for individual freedom of choice, especially on religious matters, formal *political* equality before the law and nascent *social* demands for autonomy, expressed as maximal utility in relation to privatized wealth.

Like Dumont and Gauchet, Habermas examines modernization in historical terms, as the social product of conditions in and thorough which persons came to understand themselves as autonomous individuals. Habermas draws attention to how a traditional, backward-looking, confessional and obeisant Judaeo-Christianity became unsuited to societies in which capitalistic markets for goods and services and relatively complex, intellectually abstract and spatially extensive social relations had become central. For him, a 'structural transformation' took place amidst the proliferation of contract-style agreements for the economic exchange of goods and services as social reproduction shifted from royal court and cathedral to city markets and offices. As mercantile capitalism displaced the tithe- and tribute-based economy, abstract decision-making, personal autonomy and ethico-moral choice took precedence over ancestral heritage and access to religious knowledge. This shift in the locus of social production fuelled the emergence of a public sphere of private individuals who held an interest in politics but who were not directly linked with the power structures of church or state. Important for Habermas is that the dispositions and capabilities that modernizing society required of individuals – such as effort and industriousness, techno-scientific knowledge and entrepreneurial or creative talent – became new means for attaining power in society. Whereas in traditional societies power and status had been rooted in ancestral control over real property and conversance with the divine, modernization conferred power and status upon those in possession of technical knowledge and movable wealth: gold, promissory notes and, increasingly, cash money.

In this sense, the social sources of political power that had been anchored in control over landed property and religious knowledge shifted, with the advent of modernization, to spring from personal effort and 'proper' ethico-moral choices. This situation had the effect of objectifying what Dumont sees as sometimes cruel, but always just, brute natural facts. Authority came to be seen as the product of 'an involuntary automatism', whereby the natural subjective reaction to objective brute forces of nature was to make autonomous ethico-moral choices, which led one to 'success' in earthly endeavour. In these respects, Dumont sees emphases on the political value of privatized ethico-morality as a product of cultural ideology. Discussing the work of political philosopher John Locke, Dumont contends that with modernization:

> [P]olitics as such is *reduced to an adjunct of ethico-morality and economics*. Ethico-morality and economics provide, in the 'law of nature', the basis on which political society should be constituted. Thus the

> 'true and solid happiness' that [according to Locke] should be preferred by the free and rational creature is finally human order (or what remains of cosmic order) *as it appears to the individual, who is bound to think in terms of hedonism.*[23]

With the establishment of an artificialist culture, autonomous decisions taken by individuals 'in the here and now' could radically alter personal power and status in relation to others and the world. Hannah Arendt discusses in similar terms the implications of the dissipation of power centred on immovable landed property and increased importance to political relations of universal and movable wealth in conditions of modernization. For her, the possession and accumulation of movable wealth worked as one indicator of success in worldly actions, while attention to personal duty and ethico-moral calling also lifted personal status by publicly confirming the alignment of each autonomous individual with nature's laws.[24] At once, power was constituted as something 'out there' in nature and something within individual persons, which was given expression in autonomous choice. The new artificial ideology based in the application of such ethico-morally autonomous self-sovereignty privileged those possessing control of an objectified world through movable wealth over those possessing inherited entitlements and knowledge of the cosmological order. In this view, the political faction that controlled objects through movable wealth could also exert power sufficient to grasp the capacity to organize the rules for life held in common. Hence, Habermas argues that a particular 'public' were able call for political authority to legitimate itself, based in their shared experiences of a nature each had subjectively mastered by making right ethico-moral choices and forming true rational opinions. By virtue of future-oriented ethico-moral decisions aimed at achieving worldly success, a politically ascendant public sphere effectively naturalized a form of autonomous self-sovereignty that subjectively acted both over and against the objective natural order of things. For Habermas, such a particular cultural milieu of 'self-interested, property owning private... [male, educated, literate, and] autonomous individuals' acted to establish the political means for ensuring liberty from arbitrary domination and assuring mercantilist aspirations for colonial expansion:

> The fiction of a justice immanent in free commerce was what rendered plausible the conflation of *bourgeois* and *homme*, of self-interested, property-owning private people and autonomous individuals per se. Under the social conditions that translated private vices into public

> virtues...the subsumption of politics under ethico-morality was empirically conceivable. It could, in the same world of experience, unite two heterogeneous legislations without one being likely to encroach upon the other: the legislations of private people propelled by their drives as owners of commodities and simultaneously that of spiritually free human beings.[25]

In this perspective, political conditions for the emergence of a public sphere of private individuals centred on the increasing wealth and power of a political faction that was populated by 'sovereign individuals'.[26] As Habermas shows, this grouping came to monopolize social relations through control of the economic wealth central to capitalism, such that the locus of political power within society was shifted from court and cathedral to parliament and the coffee houses, in and around which sprang up the bourses. Habermas describes how politics was shaped by 'private citizens' who – buffeted by the wealth and industriousness that reinforced self-reflection on the rationality or 'naturalness' of their own actions – 'aimed at rationalizing politics in the name of...a private ethico-morality' that saw 'itself as unpolitical':

> [The public of citizens was] a sphere of public authority [that] was now casting itself loose as a forum in which the private people, come together to form a public, readied themselves to compel public authority to legitimate itself before public opinion. The *publicum* developed into the public, the *subjectum* into the (reasoning) subject, the receiver of regulations from above into the ruling authorities' adversary.[27]

'Public' politics takes place within what Gauchet sees as 'a fracture of the truth', that is, of cosmologically ordained eternity.[28] In contrast to the situation in traditional non-modern societies, where the decisions of those holding politico-religious authority are bound to maintain cohesion between eternity and social relations in the present, in conditions of modernization the artificialism that sustains individual self-sovereignty supports a situation whereby formally equal private individuals emerge as the legitimate regulators of political power. In the modern state each individual citizen, at least formally, holds a stake in the political regulation of that state and, conversely, in the state's regulation of social relations. In relation to the state, society itself becomes cut off from the cosmological order and the state-polity emerges as a self-organizing entity based in a politics that is acted out by factions which vie for control over the capacity to organize the rules for life held in common within a given territory. It is in this perspective that Dumont, Gauchet and others, such as Claude Lefort and Cornelius Castoriadis, view modern society as essentially 'self-regulating'.[29] The holistic link with eternity – and, so, with the ecosphere – is broken. The power wielded by the self-regulating state is subjected to a political will that is expressed and exerted by a contingent collection of self-sovereign individuals who, to a greater or lesser degree, grant legitimacy to those who control the levers of power.

As possibilities for confronting existing conditions as unjust were formalized as political processes in the mercantile state, Habermas finds conditions that translated 'private vices into public virtues', provided justification for efforts to transpose assertions that society might be organized unjustly into legitimate political claims against the arbitrary exercise of authority. Moreover, in the mercantile state that emerged, political power was to be exercised in public; it was not to refer to the particular wrongs committed by individuals against the cosmological order, but to be directed at the state and highlight universal injustices committed against an abstract humanity in general. That is, as 'private vices', such as self-seeking behaviour or acquisitiveness, became increasingly necessary to the reproduction of society, a politically derived Rule of Law was required to set limits and maintain social order for the common good. In this view, assertions by individuals that a situation was indeed felt to be unjust gradually took shape as claims not against individual perpetrators but against the state. Such assertions were made in the political sphere, which for Habermas defined the contours of legitimacy and authority. Indeed, for Turner, it is 'the creation of political spaces'[30] in conditions of modernization, spaces that are delineated from private spaces by virtue of the 'universal' claims considered therein, which frame subsequent struggles over citizenship and the state. In the context of the metamorphosis of the state under the influence of expanding capitalistic markets, mercantile industrial development and urbanization, such claims centred upon what Marshall defined as *civil* demands for individual freedom of choice on religious matters that over time came to encompass claims for formal *political* equality before the law and *social* claims that, for many decades, took the form of assertions that hedonism – the pious or otherwise expression of what Biro regards as the biological instinct for self-preservation – was the right of every citizen, regardless of ancestry. Indeed, the modern bourgeois citizenship that was established resolved what Marshall sees as a particular historical instance of class or group conflict, between the traditional ruling classes and the new bourgeoisie. It institutionalizes that conflict within the state around the institutional arrangements of the courts of law, parliamentary government and the administration of taxation revenues derived from capitalistic markets by bureaucratic means, as well as the freedom to choose one's own 'brand' of hedonism, be it religious or otherwise.

The corollary of the institutionalization within the state of demands that authority legitimate itself before citizens is that citizens themselves must justify to each other their claims that social relations are unjust. In the absence of a unified cosmological ordering principle, the state's self-regulation takes place through struggles wherein the proponents of particular assertions that injustice is present must justify their claims in reference to some kind of universal validity or generalizable interest, and use a vocabulary of justice that is meaningful in relation to a shared cultural ideological grammar. It is in this sense that the critical pragmatic approach can provide important insight into the establishment of citizenship and the state in conditions of modernization. As I describe it, a central concept in the critical pragmatic approach is that of *justification*. For Boltanski and Chiapello, providing a reason or justification for action is central to secular politics:

> Inasmuch as they are subject to an imperative of justification, social arrangements [as a whole] tend to incorporate reference to a kind of very general convention directed towards a common good, and claiming universal validity... What we have called the spirit of [modern] capitalism necessarily contains reference to such conventions, at least in those of its dimensions that are directed towards justice. In other words, considered from a pragmatic point of view, the spirit of capitalism assumes reference to two different logical levels. The first contains an agent capable of actions conducive to profit creation, whereas the second contains an agent equipped with a greater degree of reflexivity, who judges the actions of the first in the name of universal principles.[31]

The metaphorical term that Boltanski, Chiapello and Thévenot use to describe political space is 'the polity':

> The [polity] is modelled on the kind of operations that actors engage in during disputes with one another, when they are faced with a demand for justification. This demand for justification is inextricably linked to the possibility of [political] critique. The justification is

> necessary to back up the critique, or to answer it when it condemns the unjust character of some specific situation. [Thus,] disputes of justice always have as their object the ranking of *status* in a situation.[32]

The two-dimensional conception of the 'actor' that is used by Boltanski and Chiapello in this sense resembles Habermas' view of the public bourgeois citizen as the political representation of, on the one hand, a property-owning individual and, on the other hand, an individual specimen of humanity in general. In other words, the actors that constitute the critical pragmatists' political communities both inhabit and constitute a state and culture. Put in Habermas' terms, citizenship is the formalizing of a mode of agency that conflates *bourgeois* with *homme*. Understood in this way, I argue that the conceptual metaphor of the 'polity' can be used to point to action within the properly political sphere of the modern state, which is always encompassed by a cultural ideology and social practices that provide the basis for ranking status in a situation. By evoking the notion of *status*, Boltanski, Thévenot and Chiapello follow somewhat the line of argument that has been developed by Michel Villey and Michael Walzer. In light of this work, they understand status in Aristotelian causative terms, such that the requirement for justice is reducible to a requirement for equality insofar as 'defining a relationship as just or unjust ... presupposes a definition of what the value of things and persons consists in, a scale of values that requires clarification in the event of dispute'.[33] That is to say, Boltanski, Thévenot and Chiapello are not referring to a quasi-psychological conception of individual status and status-seeking, as they suggest do Thorstein Veblen and, to some extent, Pierre Bourdieu. Rather, they see status as invoking 'principles of equivalence that have a claim to validity in a society at a given point in time [and] are orientated ... towards universal validity'.[34] That is, the status in question is not that of a group or individual but that of the particular political value that, represented by an actor, discourse or object, is at stake in relation to a prevailing test *of* status. In secular, self-regulating societies, where hierarchy is obscured by the formal equality that Dumont sees as the artificial principle of equalitarianism, citizens are bound to defend their interests in a political sphere that is 'encompassed' by that cultural ideological definition of what the value of things and persons consists in.

To briefly reiterate, citizenship represents a set of statuses, practices and discourses within the context of the state *and* something two-dimensionally ideological. In this second sense, citizenship represents both a potential basis for the *political* ideological 'distortion and dissimulation' of social relations and a *cultural* ideological 'whole set of representations that facilitate social integration and identity preservation'.[35] That is, citizenship can be understood not merely as a legal, political status but as a cultural way of life. Citizenship provides a framework for compromise over the legitimacy of the norms and rules that apply in a society, while also informing what it means to take part in reproducing society. Citizenship can therefore be seen as the contingent historical product of relatively peaceful struggle between social actors in the context of a particular culture and within a given state.

Conversely, citizens' assertions that injustice is present can, when successful – in the case of bourgeois citizenship, where *liberty* came to be regarded as something 'unjustly' constrained by religious orthodoxy or aristocratic privilege – be seen to shape the vocabulary of political debate. Politics, in conditions of modernization, always threatens to fragment cultural ideology by exposing as illegitimate, and so of low status, established norms and rules. Given that different actors hold different and often incompatible ideas and beliefs about how best to organize the rules for life held in common, political tensions between different groups of actors become visible when different actors portray particular values as legitimate or illegitimate. In this view, an illegitimate claim that injustice is present distorts or dissimulates the social integration that cultural ideology facilitates. A compromise arises when different conceptions of how best to organize the rules for life held in common gain support against the backdrop of a prevailing cultural ideology and social conditions, without 'clarify[ing] the principle upon which ... agreement is grounded'.[36] In situations characterized by compromise, attempts to reinforce one or another political position, or to map out an alternative, are marginalized.

This perspective points concretely to an understanding of the political sphere as the sphere of action in which the rules for life held in common are clarified and debated in view of gaining purchase upon the authority that holds the capacity to implement them, that is, the state, and against the backdrop of a prevailing cultural ideology. Such a view I believe concurs with Dryzek's political definition of the state as the 'set of individuals and organizations legally authorized to make binding decisions for a society' that is also, and importantly, subjected to the structural constraints of capitalistic economic and democratic legitimation *imperatives* that must be fulfilled 'to secure longevity and stability' of the state.[37] Moreover, it also implies a need to discuss how and why one set of norms and rules might be accepted as legitimate by a public while another is not. In short, it is within the political sphere of the state that citizens justify their assertions that a particular claim or set of claims does, in fact, reflect justice. This is because citizenly assertions that injustice is present arise in the context of a framing cultural ideology and ongoing practices of social reproduction.

In this view, 'the possibility of critique' is a product of the secularization that coincides with the modernizing of cultural ideology and the emergence of the citizenship-based nation-state. The secular modern state arbitrates compromise in the case of competing assertions that a particular set of conditions and arrangements are or are not legitimately justifiable. The state is in this view the institution charged with mediating relations between the holders of different political ideological positions and with operationalizing their claims against the backdrop of a prevailing cultural ideology, against the backdrop of or under the constraint of certain structural imperatives. This mediation must, if it is to be regarded as legitimate by citizens, take place in the political sphere that, in democratic contexts, is open to them, potentially at least, as participants. This legitimacy depends on shared cultural ideological representations that privilege the equality, freedom and autonomy of the individuals who constitute the state as a political entity. Hence, the state in this view is geared not only to defending certain functional imperatives that arise from the presence of historically entrenched structural constraints but simultaneously to upholding some notion of justice, lest that state be regarded as illegitimate. It is both structural imperatives and this status as arbiter of compromises amongst citizens that shape prevailing social relations within the state. On the one hand, the imperatives to 'keep domestic peace, respond to external threats, prevent capital flight and raise revenues'[38] constitute structures of support for what Marxists and post-Marxists call 'capitalistic accumulation' or 'growth'. On the other hand, the metamorphosis of the biological necessity for self-preservation into the compulsion to act in self-interested ways frames a political structure that is premised upon the engagement in social relations of free and equal, autonomous individuals. With the advent of the secular self-regulating state, possibilities for confronting existing conditions as unjust are formalized around a cultural ideology that values or grants high status to equality, freedom and autonomy. The state must therefore balance what amount to culturally distinct demands for justice in the context of structural imperatives.

In Marshallian terms, this politicizing of the authority to organize the rules for life held in common means, however, that the state is structurally compelled to promote particular kinds of social relations that are based upon the reproduction of inequality. In this perspective, I follow Gauchet in recognizing that the authority of the state – which arises in traditional societies from the unquestioned authority of myth and tradition – arises in conditions of modernization around the institutionalizing of arbitration or mediation over sometimes latent but oftentimes explicit 'compromise on the social bond'.[39] The state mediates differences that arise in relation to conceptions of the status of the social whole, expressed in cultural ideological principles or motifs, in relation to that of the individuals who constitute it, and who debate questions of power in the political sphere as free and equal, autonomous individuals. It is in this sense that Dumont argues for an approach to society, the state and individuals that is premised upon interpreting the encompassment by the whole of its contrary, an approach which requires seeing the particular in terms of its relationship with the whole, in a way that defines the qualities of both. The state can be seen to act as the broker of such a compromise on the social bond, but only when tensions between different political claims represent a challenge to the functioning of imperatives of state. Unless political disputes cross this horizon, they remain disputes over cultural meaning. Such disputes frequently but do not necessarily boil over into the political sphere. In this view, the essentially political problem that citizenship unleashes is a requirement for compromise between different political voices within the institutionalized political community of the state. Such compromises are brokered by the state, which aims to balance collective obligation with individual autonomy, subject to the constraints exerted by cultural ideology, insofar as the legitimation by citizens of the state rests in shared cultural values, norms and rules, and by structural imperatives to maintain external security, internal order, to raise revenue, promote accumulation and, indeed, foster legitimacy. As Turner suggests, the historical growth of civil, political and social citizenship has been typically the outcome of violence or threats of violence, which 'bring[s] the state into the social arena as a stabilizer of the social system'.[40] The state can in this sense be regarded as the secular institution that citizens, acting as 'property-owning individuals' *and* 'specimens of humanity in general', charge with the task of maintaining the social bond by arbitrating between calls for individual autonomy and collective obligation.

2.3 Bourgeois citizenship and the test of nature

What is important about the emergence of modernization as a social-historical process is the way that individuals' lives came to be stretched within a competitive market-centred social arena that demanded personal detachment and a calculative and acquisitive disposition, as well as coordinated and cooperative interaction. Habermas recognizes that the establishment of capitalistic markets especially, but also of abstract legal principles and technological and administrative realms requiring 'instrumental' rationality, created an impetus for political expression by property-owning agents who also self-identified as individual specimens of humanity in general. Habermas' account thus also highlights the equal importance to modern society of a private 'sphere of intimacy',[41] one that emphasized empathetic inter-relations and the cultivation of a self-expressive subjecthood. That is, Habermas' reasoning subjects also acted as spiritually free human beings, developing what Boltanski and Thévenot see as the 'sympathy' that conversely 'helps to transform persons subject to passions into individuals who can identify with one another and thus come to terms over a market of external goods' and which Adam Smith regarded as the 'disposition of the impartial spectator'.[42]

A peculiar cultural tension characterizes the emergence of citizenship and the nation-state in conditions of modernization, one that cross-cuts the tension between reactionary and progressive political positions. The citizenship that became established was founded upon property ownership and public engagement as well as on a 'saturated and free interiority'. Habermas demonstrates how the politically successful assertion of free and equal citizenship coincided with an understanding of individualism as something that supported a 'counter-mandatory bourgeois culture of arts and letters'.[43] Seen in this perspective, while a mainstream modern public of citizens sought to establish political stasis based in natural laws turned against themselves, an alternative quasi-public that grew out of the drawing rooms of the bourgeoisie itself, the private sphere, arose to hold private property in disregard and pursue disorder through immediate experiences and sensory gratification within them. Such a countercultural movement could assert an inner-worldly sensuality that was opposed to the competitive rough and tumble of public political engagement, and an aestheticism that diverted ethico-moral autonomy into private judgements of taste and authenticity. In the mainstream 'bourgeois' register, an alien external world is overcome by means of self-discipline aimed at domination, appropriation and the establishing of hegemony through exclusion and possession. The presumption is that social norms are justified by the objective brute forces of nature. In the counter-mandatory register, a strongly felt aestheticism is oriented to achieving unity with an externalized and 'noble' nature, from which the self is overtly alienated. The counter-mandatory presumption is that 'wild' nature, in its arbitrary naturalness, justifies noble social norms.

The division that I represent here in synthesizing work by Dumont, Habermas, Boltanski, Thévenot and others can also be fruitfully discussed in terms of Meyer's and Biro's green theories of Western political culture. In light of Biro's work, the process of modernization that 'demands' the biological instinct for self-preservation be put into practice in ways that express what is universal to individuals can be understood to express not merely what is universal but what a particular group in society considers to be universal. That is, self-sovereignty marked by personal autonomy came to be granted expression in actions that treat of objective reality in rational ways. Insofar as automatic justification for claims that the world *is* in a particular way become questionable in the secular state society, a focus on citizenship helps to delineate why norms that are oriented to achieving objectives beyond self-interested actions seemed to consistently risk falling foul of positive arguments that the existing order reflects or should reflect objectively discernible laws of nature. With a focus on the establishment of citizenship in the mercantile state, it is possible to delineate what Meyer sees as the 'competing dualist and derivative' orientations that are central to Western interpretations of political culture.[44] With the political rise of the bourgeois citizen and the civil, political and social rights and duties of bourgeois citizenship, appeals to natural laws or scientific reason marginalized possibilities that politics be instituted as a collective project that aims to replicate the nobility of raw nature. Politics came to be instituted around demands that atomistic individuals negotiate their own ways within an ostensibly objective reality. Individuals, stripped of a coherent image of the social whole and set to confronting an objective nature, thus, as the sovereign holders of a particular order of interests, encounter a historically peculiar kind of freedom. Seen in these ways, the predominance of artificialism offers a conceptual schema for thinking about the principles of individual freedom, inter-personal equality and autonomy drawn from common interests that have come to characterize citizenship as a political state of being.

By way of clarifying this point, I draw upon work by Marshall Berman, who examines the political culture of modernity by delineating modern and modernist cultural sensibilities. On the basis of Berman's analysis, I argue that to express a *modern* sensibility is to represent experiences of the disruptions that modernity brings as if from within a coherent subjectivity that responds with efforts to maintain and reinforce the coherence of the self: modern culture is dualistic, situating the individual self over and against nature. Meanwhile, to express a *modernist* cultural sensibility is to represent the experience of modernity as one of seeking to 'be at home in the maelstrom' through self-exploration 'at the fringes of self-disintegration'; that is, modernist culture is holistic, such that the self is understood as but one element of the whole.[45] I contend that the sensibilities that Berman labels modern are related to the public sphere in a historically contingent way, insofar as modern cultural ideology was established as a consequence of the importance to social reproduction of Habermas' 'private individuals'. Those who held to bourgeois modern culture were more likely to participate in industrial production, mercantile trade and politics, while those holding countermandatory feelings were more likely to be marginalized or excluded from the political sphere. Henceforth, I follow the lead of both Berman and Habermas. I use 'modern culture' to refer to a way of dealing with the experience of modernization that is based in asserting control over nature and the disruptions that modernity brings to social and subjective life, while I use 'modernist culture' to point to a way of dealing with the experience of modernization that aims at achieving authenticity and self-realization within nature. While the modern cultural ideology explicitly favours positive, objective (rational) knowledge that humanity is separate from and above nature, the modernist cultural ideology favours subjective (often avowedly 'spiritual') knowledge that the self is immersed within nature.

In the terms set out by Boltanski and Thévenot, modern ideology could be seen as representing a compromise between emergent 'industrial' and 'market' orders of justification, while what I see as modernist cultural ideology might be said to represent a marginalized yet hitherto predominant 'familial' order of justification. However, I eschew the critical pragmatists' multiplex classificatory schema as unsuited to my focus upon citizenship and the state, insofar as their approach tends to leave under-theorized the political structures that are produced when actors such as citizens, who hold certain ideological values, engage in constituting social conditions that are 'real' for them. Hence, I instead emphasize the contradiction that Lefort ascribes to ideology in conditions of modernization. For Lefort, modernization creates a tension between explicit rejection of transcendent holism, based in the embrace of positivistic 'scientific' rationality, and an implicit abandonment of dualism and recrudescence of holism, which in a particular historical moment manifested as Romantic subjectivism, for example.[46] In this light, the modern bourgeois citizens who supported assertions of civil, political and limited social rights and responsibilities did so from within a dualist-artificialist cultural ideological frame of reference. Conversely, modernists asserting rights to freedom of creative expression, as well as questions of taste, authenticity and existential meaning, did so by eschewing artificialism. The 'sadness' or 'sense of loss' attributed to Romanticism implies the impossibility of a return to pre-modern holism. Modernism thus supports the assertion of self-realization 'amidst the maelstrom' in ways that partially contradict political claims for the Rule of Law, political representation and maximal utility that were supported by modern bourgeois citizens. Self-expression, aesthetic sensitivities and inner-worldliness were – along with the allegedly 'feminine' virtues of domestic harmony, nurturance and caring – relegated to a politically inconsequential private sphere. The critical freedom that was afforded to citizenship, as the political category of the individual in conditions of modernization, consistently put modernism at risk of marginalization from the political sphere as apolitical Romanticism, abstract Utopianism or ossified and parochial Mannerism. That is, modernist cultural ideology and the challenges to artificialism that it posed were channelled into the apolitical spheres of aesthetic production, the subjective existential realm and, as Arendt and others argue, the domestic sphere of private life.

In the transformation to the modern mercantile state that Habermas examines, principles of freedom, equality and autonomy were established at the centre of the social bond, displacing hitherto cosmologically ordained Truth, which was revealed in adherence to tradition, fealty to ancestors and privileged access to knowledge or power. In this view, key parties to politics also shifted, from aristocracy and clergy to the mercantile bourgeoisie and 'politicians', charged with petitioning the state as citizens' institutional representatives. As such, with the expansion of mercantile modernization, the backdrop to political debate shifted. Assertions that injustice is present no longer referred to a timeless eternal cosmology. In modernity, such assertions refer to a shared secular cultural ideology that had become anchored in ideal representations of the free and equal, autonomous citizen: high status in the mercantile state accrued to those who were free to act in newly expanded capitalist markets and, moreover, who acted in rational ways and made the right ethico-moral choices and, so, subjugated nature. In this view, what Marshallian analysis sees as the establishment of parliament, the Rule of Law and first religious then existential freedom provides the defining features of bourgeois citizenship.[47] Meanwhile, and importantly, it was capitalistic markets that increasingly brokered cultural differences, through the development of luxury goods and services as 'fashionable' commodities, for example, while many domestic duties, such as childcare and cleaning, were drawn out of the familial realm as commercial relationships between property owners and 'hired help' disrupted traditional hierarchical relations between master and servant. It is in this sense that Boltanski and Thévenot's argument that debate between the holders of different political ideological 'justifications', when sharing a cultural ideological grammar, can tend to coalesce in the establishment of a new test that, following Dumont and Gauchet, established a new principle at the centre of the compromise on the social bond. To reiterate, the key criterion for identifying a shift or transformation from 'compromise' to 'test' is that certain political ideological commitments are abandoned and observable is the emergence of new criteria for justifying that a particular course of action or discourse contributes to the common good. While a compromise arises when different conceptions of how best to organize the rules for achieving the common good gain support within a single polity, a new test is established when the product is a new cultural ideological grammar. What is at stake in a test is 'the type of strength that is involved in a specific test and [that it] does not call upon the use of any other kind of strength. All in all, a legitimate test must always test something that has been defined, presenting itself as a *test of something*'.[48] When actors represent particular practices and discourses as worthy or unworthy contributions to political debate by representing them in relation to a different or incompatible political value, a compromise may crumble and a new test emerge.

In light of this point, eschewing Boltanski and Thévenot's multiplex classificatory schema, in favour of distinguishing between modern and modernist cultural ideological forms, also raises a need to distinguish these in relation to *reactionary* and *progressive* political ideological positions. That is, in the absence of an eternal and totalizing cosmological order of things, societies and the political communities charged with organizing the rules for life held in common within them can be understood as cleaving between those who benefit and those who lose from the current order. This is the basic distinction that is said to have emerged in the French Revolution. Building on this distinction, it is tempting to regard what I am defining as modern cultural ideology as being the preserve of the political right, such that reactionary efforts to preserve existing structures of privilege based in claims to be upholding tradition, particularism over universalism and essentialism over contingency[49] can readily be cast in terms of subjugating nature and others and appealing to rational means for enabling control over wild nature and unruly individuals. In the same view, what I am calling modernism may be regarded as the defining feature of the politically progressive left, such that efforts to reduce social inequality and promote the freedom and autonomy of all citizens are readily cast in terms of efforts to universalize possibilities for self-realization. However, complicating this picture is Dumont's own perspective, which views the reactionary political position as holistic, insofar as it seeks to establish an orderly hierarchy, while progressive politics expresses dualism because it champions the rights of the individual over the social whole.[50] Both views appear problematic given that the reactionary political position has 'always been split between attachment to a hierarchical, organic collectivism, whether traditional or, as with Fascism, revolutionary and to a [libertarian] entrepreneurial, free-market capitalism that proclaims equal property rights and equality of opportunity'.[51] Similarly, both views seem problematic given that the progressive political position has since the 1790s been at various times associated with desires for interminable revolutionary chaos and for bureaucratic administrative *dirigisme*.

Several theorists offer a way through this apparent conceptual quagmire. For Norberto Bobbio, the essential difference between the progressive and reactionary political positions is that the former is committed to justifying equality and the latter to justifying inequality. For Bobbio, 'no proposal for redistribution can fail to respond to…three questions: between whom? Of what? And on the basis of which criteria?'[52] However, while dealing with the question in terms of distributive justice, Bobbio's conceptualization of the distinction is not broad enough to encompass non-distributive questions relating to the task of defining who is a legitimate political actor or who counts as a legitimate political participant, that is, who is legitimately to experience all of the rights and express all of the duties of citizenship. Steven Lukes extends Bobbio's argument by conceptualizing the distinction from a point of view that sees the progressive political position as being grounded by 'a political ideal of equal citizenship, where all have equal civil rights that are independent of their capacities, achievements, circumstances and ascribed identities, so that government represents their interests on an equal basis; and [a] social ideal of conceiving "society", including the economy, as a co-operative order in which all are treated as equals, with equal standing or status'.[53] For Ted Honderich, it is simply the justification of selfishness that defines the 'essence' of reactionary political positions, such that 'conservatives' aim to preserve tradition, order or hierarchy on the basis of self-interest.[54] Combining Bobbio's,

Lukes' and Honderich's theses, I contend that the essential difference between progressive and reactionary political ideological positions, when understood in terms of the possibility that either may arise in the context of modern or modernist cultural ideologies, is that the former entails accepting as the primary justification for ethico-moral decision-making the interests of others as being equivalent to one's own, while the latter accepts as the primary justification for ethico-moral decisions only the subjective perspective of narrow self-interest. In this view, both modern and modernist, that is, dualistic and holistic, cultural ideologies can support both left and right political ideologies. Cultural dualism can encompass contrasting political positions that seek to justify the subjugation of nature with the aim of protecting existing subjective privileges or extending equality, freedom and autonomy to all citizens. Similarly, holism can encompass contrasting political positions that seek to justify a 'nihilism of the happy few' or the extension of conditions that support self-realization 'amidst the maelstrom' to all citizens. In this view, a test becomes visible when political tensions between citizens, sharing in common a cultural ideological grammar of representations, demand that the state or similarly empowered political institution act as arbiter of the compromise on the social bond.

At this point, it is necessary to note that the ostensibly critical pragmatic and realist approach that I apply here resembles somewhat Foucauldian and post-Foucauldian 'governmentality' theories. However, the critical pragmatic realist approach is designed to avoid a core problem in the popular post-Foucauldian approaches when applied to questions of citizenship, the state and ideology. That is, that governmentality theory tends to flatten the different levels at which people share understanding or act in consensual ways, and tends to level the difficulties or meaning that living socially raises for its members. That is, in a governmentality perspective, 'everything is political'. Moreover, while knowledge is subjected to interests and manipulation and, as such, can be seen to reveal and sustain power, 'if we understand the problem in those terms, we leave aside the more profound relationship between political and [cultural] representation, a relation that cannot be reduced to manipulation strategies'.[55] In short, governmentality theory provides an important purview over the power dynamics behind social relations, but offers less insight into the functional and constitutive dimensions of prevailing discourses. The epistemological limitations of governmentality theory mean that it does not deal adequately with how political representations may exploit, contribute to and transform 'reality'. A critical pragmatic yet realist approach thus sheds analytic light upon what motivates actors to justify the representations that they make and how, on what terms, these might be recognizable as assertions that injustice is present. As such, I treat politics not merely as the expression of 'capillary' biopolitics, as in governmentality theory, but in terms of a prevailing grammar of representations that encompasses different – progressive or reactionary – political ideological positions.

Returning to the case at hand, encompassed by cultural ideological dualism, the mercantile state was called upon to arbitrate on a test that delivered high status to free and equal, autonomous individuals who, making the right ethico-moral choices, subjugated an objectified and externalized nature. In Habermas' view, the state is, by virtue of its grounding in an a dualist 'equalitarian' cultural ideology and its need to fulfil structural imperatives, 'inherently reactionary'.[56] The state upholds a cultural ideology that defines shared representations yet is the facilitator of systemic social processes that promote particular interests, which are difficult to dismantle or re-organize without dismantling or reorganizing society itself. That is, state imperatives 'work' for many citizens, whose empowerment and privilege depend upon historically entrenched social relations. Moreover, in the narrative that I present here, what Marshall defines as claims for civil and political citizenship were largely responded to by the state, while claims for social rights and responsibilities either were dealt with as questions of personal conviction or went largely unheeded. That is, in the mercantile state, certain progressive social claims remained the province of marginalized intellectual and revolutionary progressive movements. Hence, the arrangements of citizenship that were established centred on the bourgeois institutions of the civil law courts and political parliamentary representation,[57] while 'consumption', education and healthcare were left to the private sphere.

Hence, I have examined the emergence of modern citizenship within the territorial mercantile state by drawing on the analytic categories of prevailing social *conditions* and the nature of the *assertions* of citizenship claims, and the *responses* of the state and, by extension, capitalistic markets, insofar as these are allied with the state. I have also discussed the *types* of social and political participation that arose in these contexts to centre on a free, equal and autonomous individual who engages in the economic struggle on legal terms and on the *content* of bourgeois citizenship rights to non-intervention in individual affairs and responsibilities to obey the law, including controversially for bourgeois citizens, the responsibility to pay taxes to sustain the mercantile state. I have examined these by focusing on the establishment of the institutional *arrangements* of 'bourgeois' citizenship that centred upon mercantile capitalism, limited-franchise democracy and the impartial court of law. My argument has been based in the assertion that modern emphases on productivity, self-discipline and social conformity – in contrast to modernist emphases on unity with a 'wild' natural world – came to define the politically legitimate means for negotiating social order from the 17th and 18th centuries into the 19th century and onwards. I have proposed the view that the state and its institutions were established as a politically self-regulating entity that, in and around the institutional arrangements of citizenship, is set against an objectified environment rather than confined within it. In this view, the establishment of modern citizenship is explicitly bound up with the development of the classical state imperatives and the 'test' of nature. That is, with the support of some citizens, a politically legitimate authority – the mercantile state – could promote external order and maintain internal order by levying taxation and, in so doing, foster ever-greater capitalistic accumulation based on the exploitation of 'natural' and 'human' resources.

In this view, the establishment of the classical imperatives of state in fact enforced the marginalization of modernist sensibilities as private concerns. That is, in conditions of modernization, assertions that injustice is present are not always and necessarily of the political sphere. Insofar as they do not attract the intervention of the state with the aim of regulating the relations wherein claims of injustice arise, assertions that injustice is present may be purely cultural. In short, the presence of a test at the centre of the compromise on the social bond does not necessarily sustain political obligations. Equality, freedom and autonomy are as much aspects of a prevailing cultural ideology as they are the focus for political ideological attempts to justify a particular discourse or action. The division between the political and private spheres is in this sense structurally and institutionally entrenched as a historical condition of the modernization of social life. In the perspective that I develop here, tensions between modern and modernist cultural ideologies may shape cultural differences, yet do not necessarily or automatically shape political struggles. Until, that is, they erupt publicly as political claims in view of the state and, following the state theorists, impact upon the contents and arrangements of citizenship and the functioning of state imperatives.

By conceptualizing the split between public and private spheres in terms of differences between modern dualist and modernist artificialist responses to modernization, I set the stage for discussion in the following chapters to address what I will argue is the historical recrudescence of modernism, as a cultural ideological frame for both reactionary and progressive political positions. This recrudescent modernism or holism has had the effect of legitimizing authenticity, self-expression and self-realization as genuinely important political objectives, yet does not necessarily legitimize either political ideological position. The key impetus for this argument lies in recognizing the emergence of structural and institutional conditions that meant that modernism came, by the late 20th century, to be accorded increased importance in politics. As I will argue, with the advent of what political theorists saw as concurrent legitimation and economic-industrial crises in the 1970s, modernist cultural ideological sensibilities began to shape political assertions by citizens that injustice is present.

3
Challenging Modern Artificialism

3.1 Social citizenship and the test of wealth

By the late 19th and early 20th centuries, the unprecedented expansion of industrial production increased the importance to social reproduction of manufacturing workers as a social class. Increased global inter-state competition for markets and resources, as well as concerns over military security, created a sense of urgency around questions of working and social conditions in general. The state and economy required a compromise that would secure industrial and political peace. Meanwhile, the working class itself began to organize with the aim of achieving 'full' citizenship rights, up to and including a just share of the profits of industrial development. It is in this context that a socialistic critique of industrial society and the social claims of citizenship were pushed to the centre of the political sphere. If we once more pick up the critical pragmatists' line of argument, what becomes apparent in conditions of industrialization is a shift in the way that political compromise over the social bond is enframed. The bourgeois citizenship-based conceptions of freedom, equality and autonomy that had been central to the mercantile state were, with the expansion of industrialization, partially eclipsed by the emergence of cultural and political tensions over the redistribution of economic wealth amongst formally free, equal and autonomous citizens. Indeed, it is only in the 20th century that the balance of what Marshall saw as the permanent political tension within the state, between the need for economic profitability, the taxation requirements of maintaining internal order and territorial integrity, on the one hand, and the rights of citizens to at least some degree of collective provision of social rights, on the other hand, shifts the focus of politics from universalizing civil and political claims towards an emphasis upon social claims.

While civil and political questions of justice were not ignored or excluded from politics, this concentration on social claims emerged as a consequence of the increased importance to the task of maintaining the social bond of the industrial working class, who suffered a distinct form of injustice directly proportional to the benefits of those owning the means of production. The interests of the organized working class lay in obtaining a greater share of the wealth of industrialization, such that organized labour justified its specific claims – for higher wages and access to pensions, education and healthcare, for example – in terms of the universalization of social rights and responsibilities. In Marshallian terms, just as the political antagonisms that had surrounded mercantile concerns with civil rights and political representation were neutralized by the institutionalizing within the state of arrangements that facilitated the Rule of Law and representative government, the judicial system and a limited-franchise parliament, the workers' movement sought to have institutionalized certain arrangements that would sustain social rights and ensure obligations to fulfil certain social responsibilities. With the increased predominance of industrial capitalism, the relationship between citizenship and the state came to be defined, by the 20th century, in terms of the mutual exclusivity of working-class demands for material equality between individuals within a political entity, the state, which was itself reproduced by virtue of an economic system heavily reliant upon material rather than political inequality. In this view, the emergence of the organized working class or labour movement pushed questions of redistribution into the political sphere as social questions, such that what Turner identifies as the key sociological problem that citizenship unleashes in the 20th century turns upon a requirement for negotiation over 'solidarity and scarcity' within a single state.[1]

Hence, Marshallian accounts of the transformation of citizenship in the early 20th century describe how experiences of large-scale societal industrialization prompted new assertions of citizenship rights and responsibilities. Modernization had massively and irrevocably disrupted the social life of the bulk of the population such that by the 19th century the close geographical proximity of working-class communities to the mass-industrial sites of their employment had fostered a 'consciousness' and political organization around labour issues, which was central to the establishment of a presence within the polity.[2] The increasingly organized labour movement developed what had been established as civil rights to legal protection from the arbitrary exercise of sovereign power and political rights to representation into claims for social rights to a share in the economic wealth that industrial capitalism was producing. Initially, such assertions of social rights took the form of wage regulation, industrial safety, working hours and pensions and calls for state sponsorship of institutional arrangements in support of these. Marshall and Turner also describe how citizenship responsibilities underwent a transformation in conditions of industrialization. Although most often seen as providing an account of citizen rights, Marshall also describes the establishment of citizenly civil responsibilities, to obey the Rule of Law and contribute taxation revenues, and contends that the establishment of universal franchise brought with it political responsibilities to maintain basic understanding of political affairs.[3] In this perspective, complementing assertions of social rights were responsibilities to undertake 'full-time' employment and, arguably, to accept military conscription and eschew support for communistic- in favour of liberal- or social-democracy. The political power of the organized labour movement was derived from its position as the representative of a working class that, by virtue of its capacity to withdraw labour from industrial production, could threaten state efforts to maintain internal order and economic growth and, to some extent, external security.

In some states, such as Germany, rights to social welfare were instituted 'from the top down', in response to fears that civil unrest by working-class movements could trigger a communistic revolution or severely undermine industrial capacity and weaken the state to such threats from the outside. Then following the First World War, and in view of the threat that political upheaval in Eastern Europe might spread to Western Europe and beyond, pressure upon states and markets to respond adequately to working-class demands was increased. Although a number of historical events combined to effect the situation in Eastern and Central European states, social claims for greater regulation and control of economic activity eventually spilled over into totalitarianism there, amidst the economic downturn that began in the 1930s. This breakdown of the peaceable compromise on the social bond took place in those states most deeply affected by the internationalization of markets (Germany) and the modernization of productive capacity in the West (Russia).[4] Conversely, the massive industrial expansion of the US economy and opening up of 'the West' weakened the power of the organized working class, such that social claims in that nation were less successful than those developed in

Western Europe and, owing possibly to 'the tyranny of distance' and relatively equal distribution of wealth already in place, Oceania.[5]

As a generalizing claim, however, individual rights to equality came to privilege a definition of justice that built upon struggles over 'bourgeois' rights and responsibilities and expanded these to include a 'social' redistributive definition of justice based in the state-administered transfer of economic goods and services. In this perspective, civil rights to the juridical regularization of affairs came under increasing pressure to include industrial disputation, while political claims were transformed from minimal assertions of adult male franchise into claims for full adult franchise, albeit, as Turner notes, with the exclusion of the original inhabitants of the settler societies.[6] Meanwhile, social claims were extended into claims for the rights to access the 'full social heritage' of the state, which included but was not limited to support for retired and unemployed workers and universal access to education and healthcare. In short, the form of the citizenship claims that arose meant that redistribution increasingly took the form of rights to employment, education and health, while individual responsibilities were extended to encompass individual commitments to employment-based participation within a productive, market-oriented nation-building state. In the context of the widespread impacts of the economic downturn of the 1930s and war of 1939–45, social citizenship claims were firmly established in the political sphere as the key axis of 20th-century citizenship.

Indeed, Valdivielso extends Marshall's point when proposing that social citizenship claims 'were the contingent and concrete resolution of the tension between different elements in the modern notion of citizenship and especially in the notions of justice and equality' that had been formalized around 'bourgeois' civil and political rights and responsibilities. For Valdivielso, social citizenship was the 'result of concrete social struggles and class conflict, [such that] the modern ideals of freedom, justice and solidarity were taken one step further to transform the similarly modern ideals of individual, property and market' along redistributive lines.[7] That is, the state, as the legitimate authority charged with the task of organizing the rules for life held in common, actively undertook to broker a compromise on political debates over justice and how to implement it. In particular, social claims that were defined in terms of negative liberty – expressing what Dumont sees as Lockean hedonism – were expanded to encompass positive liberties, insofar as these could be defined as inequalities produced by the capitalist economic system. The issue of redistribution emphasized citizen rights to employment or unemployment 'relief' and duties to support the Cold War military-industrial complex, over the bourgeois compromise that had defined bourgeois citizenship in terms of personal effort and 'proper' ethico-moral choices in support of the imperial pretensions of the mercantile state.[8] This compromise – analytically speaking, defining the status of the social whole in relation to the individuals who constituted it – in turn shaped the cultural ideological backdrop to the social citizenship that encompassed the political sphere from the 1940s and 1950s onwards. The types of social and political participation that were central to social citizenship, those of the worker-consumer, and the extension of the content of the bourgeois civil and political rights and responsibilities to encompass employment relations and full adult franchise were anchored by political debates over the institutional arrangements of the redistributive industrial welfare state.

This perspective on the welfare state compromise expands Giddens' argument that 'there is always a tension between state and economy; the separation of the two spheres always involves at the same time a mutual dependence', such that 'the rise of the modern welfare state must be understood...in terms of such mechanisms of the mutual realignment of polity and economy'.[9] However, in light of the view that I develop here, Giddens tends to conflate somewhat the nexus of the state with the economy and the polity. In contrast, I argue that given the historical development of the state imperative to support and promote accumulation in the context of the empowerment afforded to the working class by industrialization, the 'alignment' that takes place with the establishment of social citizenship and the welfare state is between the polity, on the one hand, and the state and the economy, on the other hand. The capitalistic market economy indeed exists in a symbiotic but not necessarily unitary relationship with the state. The state brokers a compromise more or less on behalf of itself and those holding power over the means of production: the state and the economy necessarily seek alignment, whereas the polity and the state-economy do not necessarily do so. Of course, many influential groups within the broader polity did seek to destabilize the state and the economy with the aim of re-creating society along collectivist or racialist lines in the mid-20th century. However, the most politically effective assertions that injustice was present were in general oriented by the aim of establishing economic-distributive rights within relatively regulated, market-based liberal-democratic states.[10] In this situation, institutional responses to assertions of social citizenship took the form of efforts to maintain the integrity of the state and market economy in the face of pressure from a politically enfranchised and economically powerful polity of social citizens, organized as a social movement. Instituted around three major institutions – nation-state government, commercial enterprise and organized labour – the welfare state used Keynesian policies to develop large-scale infrastructure projects, including the military-industrial complex, and to raise loosely defined 'living standards' for all citizens. In this perspective, what political economists such as Robert Brenner call the economic Long Upturn[11] brought with it militarization and nation-building efforts as well as a proliferation of relatively available commoditized goods and services, nuclear-familial housing, full-time employment and the social security system, amongst other things. Social citizenship and the institutional arrangements of the welfare state accompanied what Leslie Sklair suggests was a 'qualitatively new globalizing phase in the 1960s... [when] for the first time in human history, the dominant economic system, capitalism, was sufficiently productive to provide a basic package of material possessions and services to almost everyone in the First World'.[12]

It is in the context of citizenly demands for increased political regulation of the economy that the redistributive compromise over the social bond emerged as a relatively stable platform for social reproduction. In functionalist terms, the state imperatives of external security, internal order, taxation and support for accumulation were reinforced by capitalistic industrialization. Demands that the material benefits of economic growth should be redistributed to all on the basis of citizenship were in this view made possible by citizenly assertion of pressure upon the state as the political authority charged with organizing the rules for life held in common. Conversely, the state itself, in its 'symbiotic' relationship with the market, was bound to respond to the political power exerted by the organized labour movement. In contrast to the situation in the Eastern Bloc, where citizenship and the polity were subordinated to the requirements of the state and economy, the redistributive social compromise in the West was the product of a structural alignment between citizens, insofar as majority interests were represented by organized labour movements, the state, bound by structural imperatives, and capitalism, which itself was bound by an imperative of continual industrial expansion. In practical terms, the key point is that claims for just wages, working hours and a 'safety net' of welfare benefits could gain purchase upon the state-economy because the citizenly bearers of such claims, industrial workers, were central to social reproduction and the political community.

When viewed as a contingent resolution to tensions between the polity and the state and economy, political debate over social citizenship had two effects. First, it formalized debate between competing progressive and reactionary political movements. Across the West in the mid-20th century, in the context of the geopolitical Cold War and with the support of big business and organized labour, major political parties responded to social citizens in particular ways. In general, the parties claimed that they could do a better job of fostering economic growth, sustaining stable full-time (largely white, adult male) employment, providing housing, health and education and, importantly, expanding access to the leisure and domestic labour-saving consumer goods and services that the industrial economy was producing. To different degrees, both progressive and reactionary parties in the West saw the need to guarantee the territorial integrity of the state as a whole and the social security of individuals in society. In this view, those promoting a progressive position on the social bond demanded positive redistribution of material wealth across social classes, while those promoting a reactionary position viewed redistribution as a secondary concern and called upon the state to instead promote industrial development in order to facilitate negative freedoms to obtain an economic share in the gross national product, primarily through participation in employment. Second, a side effect of political debate over the contents and arrangements of social citizenship was the marginalizing of extreme political positions. During the Cold War-Long Upturn, politically engaged Leftist advocates of progressivism and Rightist advocates of conservatism settled upon a 'redistributive' or 'welfare state' compromise between collective-paternal and capitalist-nationalist policies. This compromise remained unstable, however, as neither of the opposing positions was substantially altered. Throughout the mid-20th century, Social Democrats supported collective welfarist ideals of justice, while Conservatives supported self-help atomism. Both did so from within a cultural ideology that, against the backdrop of the industrial state, had expanded from representations based in ideals of a free, equal and autonomous individual to incorporate the positive redistribution of economic wealth. That is, freedom, equality and autonomy came to be defined politically in terms of economic-redistributive questions.

As I have suggested, in compromise situations an 'intentional proclivity towards the common good' is maintained, without addressing political tensions directly: 'without [attempting] to clarify the principle upon which ... agreement is grounded'.[13] That is, in situations characterized by largely peaceable and ostensibly democratic struggle and compromise, attempts to reinforce a model of justice or to map out an alternative order are marginalized. During the Cold War-Long Upturn, those advocating extreme forms of collectivism, such as Communists, and those advocating extreme forms of atomism, such as Libertarians, were excluded from engagement within the political sphere of the liberal- or social-democratic state. In the perspective afforded by state theory, progressive and reactionary groups were engaged in efforts to align the interests of their constituents with state interests, through representations that appealed to the grammar of a Western cultural ideology that had incorporated 'social' concerns with the distribution of economic wealth. Communist and libertarian, fascist and anarchist movements of course played bit parts in shaping political debate and cultural ideological representations; however, they were unable to steer a course beyond 'accommodation' by the state. The liberal- and social-democratic state held sway and maintained its position of authority by fostering accumulation and growth, territorial and social security and, so, political legitimacy. That is, from within a constitutive perspective, a cultural ideology that privileged a principle of redistribution had become established, such that reactionary and progressive assertions that injustice was present appear to have concurred in granting status to the demonstration of collective and individual 'abundance', especially of material goods and services. This is not to argue that 'getting rich' formed the shared ethico-moral goal or ambition for all members of society. Rather, the point is that the compromise on the social bond was maintained by the state in and through its role as mediator of debates over how best to distribute the wealth produced by seemingly ever-expanding industrialization. The compromise on the social bond between individual autonomy and collective obligation became in this view subject to an economic-distributive 'test' of wealth.

3.2 Widespread contingency of choice: holism resurgent

The civil, political and social forms of rights and duties that I have described in relation to bourgeois and then social citizenship forms can be understood as grounding debates over injustice within a dualist cultural ideological frame. The types of social and political participation that flourished as a consequence of the establishment of effort to defend bourgeois and, subsequently, social citizenship are products of – just as the institutional arrangements of the Rule of Law, Parliament and the welfare state served to reinforce – the coherence of the self against the disorienting subjective experience of modernization. That is, the redistributive institutional arrangements of the welfare state and social citizenship represent a state effort of cooperation with markets and organized labour to support individuals to overcome the dislocations and disruptions of modernization. Full-time industrial employment and welfare schemes promoted self-discipline aimed at domination, appropriation and the establishing of hegemony over an external world by the self. The content of the rights and responsibilities that arose from individuals' membership of the welfare state society was premised on acceptance of such artificialism by adherents to both progressive and reactionary political ideological positions. In this view, both Left and Right sought to establish the grounds upon which society and its citizens should master the environment and the basis on which 'the spoils' of industrialization should be redistributed. For progressives, the promotion of mastery over nature was enunciated as a right to benefit to which all are entitled, thus placing a responsibility upon those who had obtained wealth to contribute to the overall good of all citizens. Meanwhile, for politically reactionary movements, the distribution of wealth was limited to particular types of citizens: those whose effort deserved to be rewarded.

In this view, social citizenship effectively marginalized modernism as a cultural ideology from the political sphere. Through the institutional arrangements of the welfare state, some aspects of the private sphere to which modernism was relegated, such as certain domestic services related to healthcare, housing and education, were politicized, insofar as the state and market actors 'agreed' to provide them in return for industrial peace and productivity. Meanwhile, those aspects of social life related more intimately to self-expression, self-realization, creativity and authenticity remained embedded in the private sphere. And, markets for consumer goods and services expanded to encompass those aspects of social life that had supported individuals' self-affirmation through embeddedness in a shared culture. This abstraction of the private sphere through commoditization contributed to a situation that Jackson Lears argues amounted to a political system 'organized to satisfy human desires…a picture of abundance based around a surfeit of mass-produced, disposable commodities'.[14] In these decades, markets and the state together promoted home ownership and private transport to citizens as sources of autonomy and independence from others. Meanwhile, markets saw themselves as respondents to the demands of social citizens, developing and promoting an increasing array of personal-use consumer goods and services.

Indeed, two deeply influential sociological shifts in the second half of the 20th century were the emergence of consumerism as a defining feature of social life and de-urbanization and suburbanization. Consumerism implies, first, political-economic reliance on demand for non-essential goods and services stimulated by markets and, especially, the knowledge or creative industries. Second, it implies that such goods and services offer to individuals a sense of commonality and identity: they serve as 'anchors of belonging' that compensate citizens for the dislocations and anxieties that are intrinsic to a highly competitive, market-oriented society. Suburbanization implies a separation of land-use patterns, whereby production and commerce remain in urban areas, while habitation is moved to the suburban fringes. Amidst industrial demand for workers, and combined with a demographically significant 'baby boom', population growth prompted large-scale private and public suburban residential development across Anglo-American societies. Such land use drew upon strong cooperation between the state, business and to some extent organized labour. The state, on behalf of its 'traditional' modern constituency, capitalist business, and its 'new' constituency, organized labour, undertook to promote the commercial development of tract housing, dormitory suburbs and so-called new towns to accommodate lower- and middle-income groups. This suburbanization has been described as creating 'doughnut' cities, whereby persons in middle and lower-middle income groups were encouraged to leave inner cities for the new developments. What has been discussed in urban theory and geography as the flight of the relatively well-off middle class from the inner cities and the large-scale development of suburbs of automobile-dependent nuclear families in the 1950s and 1960s did, in this view, have the effect of loosening bindedness to extended family, community and place while shoring up markets for personal- and domestic-use commodities, including, importantly, automobiles.[15]

As such, suburbanization, private commuting and the rise of the consumer-service economy, amongst other things, partly undermined the social cohesiveness that had been central to the organization of labour as an effective social movement and anchor-point for citizenly claims of injustice in the political sphere. At the same time, merit-based employment within bureaucratically administered government and commercial organizations brought a degree of liberation from the strictures of class and status groups. Expanded universal access to secondary, technical and tertiary education, basic primary healthcare and rights-based welfare also had the effect of weakening gender and ethnic chauvinism. For Mike Featherstone and David Harvey, amidst a plethora of commoditized goods and services in the 1960s, a generation of 'baby boomers', unfamiliar with depression or wartime shortages, gained access to disposable income and 'entered higher education in numbers higher than ever before'.[16] Distance from local community and extended family through social and geographical mobility, relative economic and personal independence and expanded access to higher education also meant to some extent that the authority wielded by 'moral custodians' and the 'local rumour mill' became an anachronistic irrelevance for many. This relative independence from others and proliferation of personal-use consumer items altered the cultural norms that shaped subjective orientation within the world; goods and services that had once been the preserve of the extremely wealthy few were made widely available, and a host of new goods and services were targeted at individual worker-consumers. And where 'extreme symbolic and practical variety [had] become a common feature of all types of commodities…self-reflection as a consequence of choosing between many alternatives [became] permanent and ubiquitous'.[17]

The combination of economic growth through industrialization with social citizenship and the welfare state ushered in what Gerhardt Schulze describes as *affluence*.[18] Schulze's argument resembles that of economist John K. Galbraith, who found that the affluent society of the 1950s and 1960s extended material benefits to large numbers of citizens on the basis of their participation in economic activities, while restricting the economic development of unprofitable persons and areas.[19] As does Galbraith, Schulze argues that affluence is not limited to personal wealth or to commodity consumption *per se* but is a generic sociological condition. Schulze contrasts affluent with non-affluent societies, where persons may possess vast wealth in relation to others in the same society, but would be simply rich individuals. Such wealth does not amount to affluence, because these societies may be strictly authoritarian, lack transparent administration or include the arbitrary use of force over citizens' lives. Jean Baudrillard also discusses Galbraith's concept in a similar way, suggesting that affluence becomes the benchmark or 'integral logic' of Western society, which as a concept does not bear comparison with riches or personal wealth in situations of scarcity or violent repression.[20]

For Schulze, the general scarcity that had characterized social relations until the 1940s and 1950s was replaced by widespread contingency of choice not only in relation to material goods but across most areas of life, for most people. Hence, affluence 'is not simply a phenomenon caused by prosperity; more generally it is a phenomenon of modernization which will not be driven away by unemployment, recession and stagnation of real income'. Schulze argues that, amidst conditions of affluence, selecting from a range of contingent possibilities displaces influencing the world as the primary subjective orientation within it. In this view, 'the stresses of getting through life' require 'situation management' and mean that self-reflection, as a consequence of choosing between many alternatives, becomes a permanent and ubiquitous condition. For some analysts, this cultural development can be misunderstood as a shift towards extreme selfishness. In contrast, Schulze argues that recognizing affluence merely implies the emergence of conditions whereby the traditional uncertainty of not knowing the means by which goods and services could be obtained is displaced by generalized uncertainty over not knowing the ends that should be aimed for. Paraphrasing psychologist Erich Fromm, Schulze argues that affluence shifts the focus of the goal of consumption from having a good or service to being one's self. In Schulze's argument, widespread affluence normalizes a shift from situation-centred to subject-centred ethico-moral discourse within society. While situation-centred ethico-morality had defined the good in terms of external conditions – money, a home, automobiles, personal-use commodities, prestige and status – the subject-centred ethico-morality that arose in the 1970s defined the good in terms of self-experience and a 'rationalization of experience' through which individuals tend to 'attempt to optimize "outside" means in relation to "inner" ends'.[21]

In this perspective, citizen assertions that injustice is present and the institutional responses to these that had led to the establishment of social citizenship can be understood as sharing the dualist grammar of a situation-centred ethico-moral discourse. The types of social and political participation and the contents of the rights and duties of social citizenship had remained politically legitimate, so long as the institutional arrangements of the welfare state were predicated upon acceptance of the state as arbiter of a compromise on the social bond that turned upon a test of wealth. A key consequence of social citizenship over time was the generalizing of conditions for a cultural shift that, through the commoditization of the means to fulfilment in private life, led to an expansion of support for modernism. The normalizing of widespread contingency of choice almost regardless of individual wealth worked to displace the situation-centred ethico-morality that had prevailed. Cultural concerns with maintaining the coherence of the self by confronting the disorienting impacts of modernization were displaced by concerns with experiencing the exhilarations of modernization, an ethico-moral stance that was hitherto limited to a bohemian artistic or countercultural fringe or else constrained by remnants of Protestantism. Affluence therefore implies a social condition in which almost unlimited contingency of choice and cultural support for self-affirmation, largely but not exclusively through commoditized goods and services, becomes central to social reproduction. Commoditization is in this sense a contingent but not a necessary condition for the self-realization that is central to modernist cultural ideology. The shift to affluence, when defined in terms of an ethico-moral shift through which widespread contingency of choice displaced existential questions of means to ends, provides an alternative view to 'post-materialist' arguments that essentialize the effects of economic wealth. This view stands in contrast to those of Blühdorn, who follows Ron Inglehart in arguing that such a shift is best defined in terms of 'post-materialism'; Theodore Roszak, who had championed the normalizing of non-conformism and desires for self-realization in the 1970s but who by the mid-1990s regarded it as a historical aberration symptomatic of the Long Upturn; or Christopher Lasch, who regarded this shift as a sign of creeping cultural 'narcissism'.[22]

In contrast, I regard such values as the legitimate consequences of increased exposure to social *and* environmental injustices. 'Affluence', including access to education and social security are sufficient sources for awareness of the unsustainable nature of social life of themselves.[23] In this perspective, the importance of whether or not subjective 'post-material values' or an objectively discernible 'post-materialist culture' are conditions of wealth appears less relevant than analytic concerns with the structural conditions that frame the transformation of citizenship and the actions of citizens who act when confronted with injustice. As such, understood in historical terms, the massive expansion of industrial production and the prevalence of social citizenship in the West had the unforeseen yet contingent impact of promoting the expression of hitherto marginalized modernist cultural ideology; almost regardless of class, race or gender, decisions over 'what is to be done' gave way to agonizing over which amongst a range of options to choose. Indeed, affluence as Schulze defines it is symptomatic of the generalizing of a modernism that supports self-realization and self-development, and is only secondarily and contingently related to 'consumerism' or post-materialism. That it was capitalism that supported such a cultural development is in this sense a historical contingency that must be dealt with analytically. In the next section, I examine the 'legitimation crisis' that arose in the 1970s as a partial consequence of the generalizing of modernist cultural ideology that I have discussed here. I develop the view that the postindustrialization that political economists see as the source of this legitimation crisis can also be understood as pushing debates over redistribution out of a position of central importance to the compromise on the social bond.

3.3 Terminal decline or democracy without freedom? The 'legitimation crisis' of the 1970s

As postindustrialization displaced industrialization within the West, both New Conservative and post-Marxist theorists sought to identify the cultural and political sources of the 'legitimation crisis' that they observed. By the 1980s, what from within both perspectives appeared to be the central strategy for satisfying the legitimation imperative of the state – expanding welfare spending – appeared in large part no longer feasible. That is, both New Conservatives and post-Marxists saw cracks appear in the state's authority *and* capacity to mediate the compromise on the social bond. Writing in the 1970s, Daniel Bell linked intensifying political and economic 'crises' with the emergence of postindustrialization across the West. Bell argued that postindustrialization took place from the 1950s and 1960s onwards, as the locus for economic activity and Western social reproduction generally shifted away from producing goods by transforming the material of the world in heavy industry and towards developing, transferring, storing, processing and applying information by managing knowledge about the world in the service sector. For Bell, the most important social and cultural condition of postindustrialization is the emergence of a technocratic or knowledge class. He argues that information-intensive knowledge industries, based on technical and professional services, and on human services, require relatively well-educated and articulate, technical, professional or, at least, semi-skilled and skilled personnel and, consequently, less unskilled manual labour. Indeed, Bell makes clear that postindustrialization does not '"displace" industrial...or agrarian society'; his claim is rather that the 'management of data and information' that increasing societal complexity necessitates merely fosters such a shift in 'social structures'.[24]

For Bell, this knowledge class displaces industrial labourers as the group most important to social reproduction in Western societies. However, he also argues that this relatively well-educated and articulate class tempers policy and provides a source of judgement that itself creates political authority, undermining but not displacing the Establishment as the major political elite. Bell's view is that a new middle class arose to displace the organized working class from a position of influence in the political sphere, while the state itself remained defined by the individuals and organizations authorized to make binding decisions for society as a whole. He makes clear that postindustrialization is predicated upon knowledge-based control and prediction of social processes, such as economic growth. For Bell, the primary institution of postindustrial societies is the university, where theoretical knowledge is codified and tested, training is conducted and policy advice is developed. Where 'the multiplying complexities of modern social and economic organization [mean that] all forms of decision-making take on a technical character', Bell finds that 'the formation of policies concerning industry and the economy devolves into the hands of [the] technical specialists' who are the new middle-class cultural elites.[25] Importantly, Bell regards the knowledge workers of the new middle class as antipathetic to the 'axial principle of economizing' around which such knowledge-based control and prediction is for him actually oriented. Bell seems to argue that, like industrialization, postindustrialization is propelled by technological development and the 'trajectory of the economic impulse' to accumulate wealth through hard work. He argues that the advent of postindustrialization marks a *caesura*, as the groups central to it embrace a hedonistic way of life whose promise is the 'voluptuous gratification' of the lineaments of desire. Bell seems to argue that the skills and dispositions for knowledge work facilitate understanding and action in the face of the failures of scientific knowledge and a kind of cultural *hubris* that unleashes the 'self-infinitizing spirit of the radical self'.[26]

Bell synthesizes theses on asceticism and acquisitiveness in the work of Weber and Werner Sombart to describe the cultural orientation central to the historical development of Western capitalistic, bureaucratic and technological industrialization. '*[A]sceticism*...the principle of the economic realm...discipline, delayed gratification in order to save and invest, and commitment to work' constantly opposes '*acquisitiveness*...hedonism, pleasure as a way of life', the principle of the cultural realm.[27] For Bell, as the knowledge class proliferates, hedonism displaces self-control and delayed gratification, purposeful behaviour in the pursuit of well-defined goals, bringing every individual into conflict with the role requirements of the technical-economic order. In postindustrial conditions, mounting cultural contradictions overload the mediation that is effected in the 'political realm, which regulates conflict [in an orderly and regularized manner] and is governed by the axial principle of equality', such that tensions between bureaucracy and equality frame the social conflicts of the day.

Bell locates the source of the legitimation 'crisis' in the tension between 'bureaucracy', that is, the state-economy nexus, and demands for equality that emanate from within the polity. Bell's approach, however, leads him to argue that the legitimation crisis accompanies a kind of ideological exhaustion that undermines state capacity to administer the accumulation imperative. For him, modernity *is* individualism; the problem is that postindustrialization unleashes a rapacious, amoral, vain and self-obsessed individualism from social, religious and cultural ideologies that once sustained the industrial order. In postindustrial society, a deeply individuated self renders what Bell sees as constraining ideologies meaningless, as unrestrained selfhood emerges as the sole point of reference point for agency. At one level, this end of ideology arises as technocratic elites achieve relative personal wealth, attain knowledge and articulateness and liberate themselves from the mannerisms of class-based societal structure. For Bell, academia, the fine arts and an exceedingly liberal polity constitute arenas in which the postindustrial class reformulates radical avant-gardism and promotes modernism as generally acceptable cultural orientations. The result is a generalized resistance to constraint and a hedonistic obsession with the inner emotions rather than the external world that, in the context of an equalitarian society, spreads beyond these circles. At another level, this end of ideology manifests in what appear through the New Conservative lens as an ever-expanding series of claims upon the welfare state. As widespread affluence and the irregular work cycles of the service sector become normalized, popular allegiance to existentialist and humanist ethics and an abandonment of labour in favour of consumption and leisure and pleasure increase.

For Bell, the postindustrial new bourgeoisie elevate bohemian aestheticism and self-indulgence to the apogee of Western culture, while the masses descend into banausic avarice and obsession with the minutiae of personal emotions and private desires. In this sense, Bell's thesis resembles those proposed by Michel Crozier and Irving Kristol, who argues that a legitimation crisis became established in these decades as the left-liberal intelligentsia came to exert undue influence over mass media and social research.[28] The problem here is that the overall argument telescopes back, from the 'crises' that are uncovered, through the arbitrarily naturalized values of the Protestant work ethic, to an *a priori* grounding in Judaeo-Christian presumptions about the fallen state of humankind and continual decline since then. What is interesting about these arguments is the authors' clear pre-analytic bias towards modern dualism. Bell, Kristol and Crozier regard support for bureaucratic order, self-discipline and delayed gratification as necessary for the economic development of Western society. These authors represent modernization as a narrative of declension towards universal proletarianization. They fail to recognize that what Schulze regards as the 'subject-centred' ethico-morality that was established as a consequence of widespread affluence has no necessary relationship to support for or opposition to economic growth. Rather, for them, in a civilization that has been built by virtue of disciplined human effort directed at subduing nature, the advent of postindustrialization elevates the status of hedonistic pleasure-seeking and atomistic selfishness, rendering all that is worthwhile dysfunctional and meaningless.[29] The source of the ridicule directed at countercultural 'peaceniks' and 'free-lovers' lay in New Conservative self-assurance about the necessity, but possibly nervousness concerning the prospects, of subduing that which is external to the self. For them, seeking to understand others or participate in 'wild nature' appears as naïve folly. In this argument, antinomian hedonism envelops society and undermines economic growth or, at least, political support for it, while eroding the will to engage in productive labour, defined as full-time employment, and destabilizing the nexus of state and the economy. This perspective tends to sentimentalize the obligations of social citizenship and modern cultural ideology, while disregarding the direct relationship of these to political criticism of markets and the state as sources of distributive injustice.

Post-Marxists such as Habermas and Giddens bring a different perspective to the social upheaval associated with postindustrialization and offer an alternative account of legitimation crisis. Central to the post-Marxist thesis is the observation that, as the heavy industries which once employed the mass of low-skilled and skilled manual labour moved offshore, the occupational groups that had once been important to economic growth were scattered across a largely non-unionized, low-wage and increasingly casualized service sector and the informal shadow economy. This dissolution of the class grouping most deeply affected by the injustices of industrialization undermined the political and economic power of the support base for welfare statism and social citizenship, which had been anchored in the organized labour movement. For Giddens, postindustrialization actually entrenches a distinction between primary and secondary employment. This distinction encouraged capitalist actors to seek alliances with the state against organized labour, the political agency of which was increasingly corralled into limited 'economistic claims' by inflationary conditions.[30] That is, the interests of organized labour in societal emancipation through the project of economic redistribution were narrowed to defending union members' wages. Meanwhile, major parties that were linked with organized labour adopted neoliberal policy platforms that sought to continually increase economic growth with the aim of maintaining aggregate 'living standards'. In Habermas' argument, the Western democratic state undergoes a fundamental re-orientation in the 1970s:

> [D]emocracy...is no longer determined by the content of a form of life that takes into account the generalizable interests of all individuals. It is now only a key for the distribution of rewards conforming to the system, that is, a regulator for the satisfaction of private interests. This [postindustrial] democracy makes possible *democracy without freedom*. [It] makes possible only compromises between ruling elites. No longer all politically consequential decisions, but only those decisions of government still *defined as political* are...subject to the precepts of democratic will-formation.[31]

In this situation, Habermas argued that postindustrialization prepared the way for the entrenching of an under-employed and unrepresented 'underclass':

> [U]nderprivileged groups are not social classes, nor do they ever even potentially represent the mass of the population. Their *disenfranchisement* and pauperization no longer coincide with *exploitation*, because the system does not live off their labour. They can represent at most a past phase of exploitation. But, they cannot through the withdrawal of cooperation attain the demands that they legitimately put forward.[32]

Initially the topic of empirical research in the United States in the 1970s, the establishment of such an underclass as a structural and institutional condition of postindustrialization also began to be discussed by British, French, German and Australian researchers in the 1980s and 1990s.[33] In this view, the decreased power of organized labour as a movement to disrupt the process of accumulation was combined with a lack of capacity on the part of its political wing to represent the interests of labour as those of society in general before the state and within the economy. The compromise on the status of the social whole in relation to that of the individuals constituting it, brokered by the state – with the aim of adjudicating debates over justice that centred on claims for a fair share of the economic wealth that industrialization had generated – was in this sense undermined. Whereas the state had undertaken to arbitrate and regulate such a compromise in terms that favoured the redistribution of economic wealth between classes, conditions of postindustrialization foster no such pressure. The pressure upon the state to legitimate itself before the polity is redirected. In short, postindustrialization led to a transformation of the nexus between the state and the economy, on the one hand, and the broader political community of citizens, on the other hand. Postindustrialization brought the decline of the industrial working class as a central force within the political sphere and, through party representation, a key actor in the state. This process undermined the state's capacity to support the conditions for compromise defined in terms of the redistribution of economic wealth. This shift had the effect of partially undoing the welfare state and undermining social citizenship.

For the post-Marxists, what had been established in the 20th century as a clearly defined democratic legitimation imperative of state had altered the terms on which the state and markets responded to citizen assertions that injustice is present. Insofar as the provision of welfare transfers out of taxation revenue supported the institutional arrangements of social citizenship, the task of maintaining internal order and civil peace through repressive security measures was partly relieved.[34] What Habermas and others such as Claus Offe and James O'Connor recognized was that, by the 1970s, the redistributive claims of social citizenship had 'overburdened' the state in its capacity to facilitate capitalistic accumulation. In particular for O'Connor, it was the increased burden that welfare expenditure placed upon taxpaying citizens themselves that was fuelling the 'crisis'. Over time, the pressure of this structural 'contradiction' between the legitimation and accumulation imperatives of state had different effects in different states.[35] On the one hand, in the liberal-democratic Anglo-American states, the welfare state compromise was weakened by the advent of a robust and aggressive form of market-oriented neoliberalism. On the other hand, in the social-democratic Continental European and Scandinavian states, the corporatist compromise on the welfare state was deepened, until the 2000s, by which time neoliberal reformism had become central to politics and policy in these states as well.

For the New Conservative theorists, the normalizing of both cultural and political challenges to the social order was the result of actions by ethico-morally unprepared or insufficiently strong individuals. It was in such terms that Bell, Crozier, Kristol and others equated the normalizing of modernism with Western decline. Hence, the New Conservatives regarded Western society as facing a crisis of decline from which it could be resuscitated only through the application of ever-greater cultural, political and, given the opportunity made available by the 'off-shoring' of productive capacity, economic discipline. Looking back upon the 1970s from the early 2000s, other such as Joseph Heath and Andrew Potter[36] cast the normalizing of modernism as the foundation of a cultural barrier that, by granting primacy to unbridled hedonism, has undermined the capacity of 'ideal' markets grounded in sensible hedonism to support a good society. In contrast, post-Marxists such as Habermas observed the consolidation of cultural assertions concerning injustice in the political claims of the new social movements, especially around environmental, nuclear-proliferation and lifestyle issues. In this view, it was the new social movements that held the potential to universalize expectations that life should be meaningful and grounded in self-reflection rather than in relation to a status group, as well as conducted in harmony with Nature and Others.[37]

However, both the New Conservative and post-Marxist theorists of the legitimation crisis attempted to rescue modern dualist cultural ideology from the effects of postindustrialization. While the New Conservative societal critique became mired in a narrative of declension that stretched back to Judaeo-Christian notions of the Fall, the post-Marxist critique lamented the dissolution of the emancipatory dimension of the dualism that had come to be associated with social- and liberal-democratic iterations of social citizenship rights and duties. They saw in the declining power of organized labour as a social force the weakening of the main tool for advancing modern conceptions of justice. In this latter view, the postindustrial fragmenting of the economic and political power of the industrial working class meant that no unitary group remained within the political sphere to petition the state and markets on the central justice issue: economic (re)distribution. This is not to argue that what the post-Marxists saw as the legitimation imperative of state was rendered irrelevant by the decline of the politically and economically empowered industrial working class, which had forced the institutional arrangements of social citizenship and the contents of its rights and responsibilities onto the political agenda. Rather, my point is that a metamorphosis of political legitimation took place in conditions of postindustrialization. For Habermas, Offe and others, the means by which the state could maintain its legitimacy was through the spending of taxation revenues on an ever-expanding (and more expensive) suite of social citizenship 'welfare' arrangements. These included welfare benefits such as unemployment payments, sickness and pension allowances, public educational and healthcare facilities. However, as the post-Marxists and others such as Pierre Rosanvallon note, the economic poverty of the socially excluded was pushed off the political agenda with the declining influence of organized labour's power to advocate on behalf of the universal social citizen.[38] The legitimation crisis of postindustrialization ushered in a cultural shift that affected what could be universalized politically as the contents of citizenship right and duties.

Part II

4
The New Citizenship, Imperatives of State and Questions of Justice

4.1 The holistic challenge to dualism

By the 1980s, what was seen by post-Marxists such as Habermas as the dissolution of the political power of a form of life that takes into account the generalizable interests of all individuals pointed to a more comprehensive transformation of the legitimation imperative of state. While 'old' social movements continued to engage in civil actions that claimed generalizable interests, these were marginalized from the political sphere itself in postindustrial conditions. Whereas the politics of legitimation had centred on *détente* between state and economic interests in security, order, accumulation and legitimation through the welfare state, on the one hand, and the view that redistributive social citizenship would eventually be universalized, on the other hand, the ideological encompassment of the compromise on the social bond shifted in the 1980s. By the 1990s, state interests in maintaining political legitimacy no longer lay in managing the compromise between collective obligation and individual autonomy by reference to the modern 'situation-centred ethico-morality' that sustained economic-redistributive policymaking. The concerns of social citizenship, as expressed by organized labour, could no longer attach themselves to the accumulation, external security or internal order imperatives, because the historically progressive focus of the legitimation imperative had been challenged. While Habermas had conceived of the distribution of rewards 'conforming to the system' as an ever-increasing burden of redistributive transfers upon the state and economy, by the 1990s the distribution of such rewards had in effect taken the form of state provision of access to consumption choices and self-directed opportunities to embrace prosperity. This shift away from a progressive politics of redistribution is recognized by Mike Davis, for example, who argues that in the United States at least, a new reactionary politics had emerged in the early 1980s around the discourses of entrepreneurial freedom from regulation, decreased social welfare and reduced taxation on individual 'consumers'.[1] Indeed, by the early 1980s, Habermas himself had accepted that modern notions of redistributive justice were themselves somewhat problematic for universalizing human equality, freedom and autonomy and had recognized the centrality to justice of opportunities for self-realization, authenticity, peaceable relations with others and harmony within nature.[2] In short, the increasing influence of what has come to be understood by the blanket term 'neoliberalism' relieved the state of its obligation to continually expand the welfare services that Habermas and Offe saw as overburdening the state in the same moment that citizens themselves had begun to reject modern culture. What emerges is a new form of legitimacy addressed by the state through provision of opportunities for self-realization through consumption of personal-use commodities, individual innovativeness and creativity.

Moreover, postindustrialization can also be understood in terms of the normalizing of cultural ideological holism through the modernist 'counter-mandatory culture' that Habermas had earlier identified as a source of alternative resistance within the mercantile state. Amidst widespread affluence and the generalizing of a 'subject-centred' ethico-morality, what had arisen in the 1960s and 1970s was a culture of dealing with the experiences of modernization by aiming to achieve personal authenticity and self-realization, a cultural orientation that stood in contrast to the modern culture of control over the disruptions that modernization brings to social and subjective life. While this shift has been described in terms of the emergence of postmodernism, I follow Habermas, Berman and others such as Terry Eagleton,[3] who contend that such sensibilities are more accurately described as modernist, insofar as the much-vaunted incredulity towards meta-narratives is more accurately understood as a historical shift through which efforts to feel 'at home in the maelstrom' of modernization were normalized. However, it is Dumont who noticed what is most interesting about the cultural transformation that took place in the aftermath of the 'legitimation crisis' of the 1970s. Writing in the late 1970s, he argued that widespread 'protest against the disruption of natural equilibriums has for the first time in public opinion placed a check on modern artificialism'.[4] While partially subsumed under the predominance of modern dualist-artificial cultural ideology over the late 19th into the

20th centuries, modernism arose to challenge the cultural *and* political predominance of dualism in the 1970s and 1980s. In postindustrial conditions, modernism gradually provided a challenge to the institutional arrangements of social citizenship in the industrial welfare state by evoking a new vocabulary for addressing the relationship between citizens and the state. That is, widespread rejection of Fordist industrial organization, Taylorist management and bureaucratic administration directly challenged the modern culture that underpinned such policies and processes. Demands for autonomy and creativity at work, the localization of control over community affairs and calls for 'free' self-expression explicitly privileged modernist ethico-moral concerns with subjective authenticity. In light of the new political culture the so-called organized capitalism of the welfare state seemed to promote psychically repressive conformity and to stifle personal creativity, self-expression and self-realization.

However, it is important to note that while the new social movements were articulating political challenges to the dualist framing of social problems in terms of economic distribution, consumer markets and creative industries began to address burgeoning countercultural movements, such as the 'sexual revolution' and more diffuse self-help and New Age spiritualist fads, which also gave expression to cultural modernism. As Habermas and others such as Regis Debray recognized, by the end of the 1970s 'the world of imagined opportunities' that modernism had brought with it began to confront considerable structural and institutional barriers to the fulfilment of political expectations.[5] This affront to the progressive, emancipatory aspect of modernism was in part a consequence of the heightened international competition between states and financialization of the economy that political economists associate with the postindustrialization.[6] That is, the establishment of globalizing neoliberalism and inter-state competition to attract investment exacerbated those institutions' demands for flexibility, privatization and market-oriented re-regulation, the development of post-Fordist 'just-in-time' production and procurement strategies and an expansion of the service, finance and creative elements of the economy in general,[7] precisely at a time when 'counter-mandatory' modernism was effectively challenging the predominance of dualism in politics and culture. As state demands for reduced social spending, employer demands for workforce flexibility and the expansion of personal-use consumer goods markets gained strength, cultural assertions that obligations to full-time employment, military service and mass consumption were sources of unjust psychic repression and stifled creativity and self-expression were gradually assimilated to a neoliberal policy agenda. However, this shift also delivered impetus to groups that could politically assert views that militarism, chauvinism, xenophobia and environmental and social harm, amongst other things, were unjust.

In this view, calls for heightened valuation of experiential authenticity and personal creativity merely *coincided*, as did increased support for social heterogeneity and awareness of ecological inter-relatedness, with the postindustrial expansion of 'consumerism' and institutional demands for workforce flexibility and innovation. Just as many of the claims associated with modern dualist culture were incorporated into the bureaucratized institutional arrangements of the welfare state and conformist duties of social citizenship, claims associated with modernism, once sidelined, were in certain important respects absorbed within postindustrial social institutions in ways that reduced their emancipatory content. For Boltanski and Chiapello, 'The price paid by [political] critique for being listened to, at least in part, is to see some of the values it has mobilized to oppose the form taken by the accumulation process being placed at the service of accumulation in accordance with a process of cultural [ideological] assimilation.'[8] It is precisely this 'real' process of assimilation that the normative theories of green citizenship fail to adequately consider when recommending the eschewal of dualism, the public-private division, territorialism and contractualism and awareness of the finiteness of ecospace. In short, these five critiques represent a political critique that is for another time. The anchoring critiques of green citizenship directly challenge the Fordist and industrial, state-centric politics of the bureaucratic, liberal-democratic, Western welfare state. They build upon norms that challenge dualist alienation from nature, mass-produced goods and services as sources of inauthenticity, full-time employment and welfarism as stifling personal spontaneity and creativity and the economic growth-oriented nation-state as reproducing structured unfairness and socio-ecological 'risks'. These critiques address an industrial, Fordist and relatively state-organized, that is, to some degree corporatist or at least Keynesian, pluralist society; an economy centred on 'full-time' waged employment in large-scale, often heavy industry; the mass consumption of mass-produced goods and services, including a unidirectional mass media; a cultural realm that privileges 'fitting in' with class and status-group expectations; an entrenched, large-scale 'top-down', technocratic and bureaucratic government; internal politics shaped by tensions between different factions vying for the spoils of industrial production through some form of welfare state 'compromise'; and external politics shaped by the Cold

War. The problem that I highlight here is that, in the Western liberal democracies at least, it is currently the postindustrial, post-Fordist, horizontal-networked politics of globalized 'governance' and the ambiguous ecomodernization of the competition state that support the injustices that have since the 1980s and 1990s become associated with unsustainable development.

Indeed, beginning in the late 1990s, sociologists of work such as Paul Heelas began to identify a broad-based shift away from 'cold and heartless industrialization' and towards a form of 'soft capitalism' that discursively reconstructed employers as benign 'caring and sharing' members of the wider community. Similar to Boltanski and Chiapello, who focus almost exclusively on the French experience, Heelas and others such as Paul Du Gay argue from their observations of Anglo-American societies that a key premise of the 'new wave' of management theory and state withdrawal from industrial relations that became prominent in the early 1990s was the assumption that individuals bring to workplaces an autonomous, creative, authentic and entrepreneurial selfhood. This 'new wave' coincides with a shift away from welfare statism and towards policies based in recognition that meritocratic equality of opportunity and scope for staff creativity and autonomy are the key to economic prosperity. Both government policy and business strategy increasingly sought to harness the psychological striving of individuals for autonomy and creativity and to channel them into the business' search for excellence in commercial practice while purporting to support individuals in the construction of an authentic identity. This transformation from Taylorist management towards an emphasis on 'workplace culture' shifted perceptions of labour away from 'an activity undertaken to meet instrumental needs', such that occupations at all levels were cast as the 'means to self-responsibility and hence self-optimization... [indeed] the new wave is explicitly aimed at everyone'. This new wave of institutional management practice and discourse extends an invitation to managers, salaried staff, waged employees, contractors and suppliers to join the corporation and the privatized state enterprise *In Search of Excellence*, while congratulating end-users and consumers for choosing to support *Best Practice*.[9]

A 'new era' had emerged, where the 'shared purpose'[10] of achieving excellence displaces class and industrial antagonisms. By the 2000s, social researchers such as Barbara Ehrenreich and Juliet Schor began to see, in ever-further employer demands for flexibility and creativity and partial abandonment of the welfare state, an increasingly atomistic workplace culture and an inconstant and precarious way of life. Similar are the concerns expressed by Madeleine Bunting over the cultural effects of workplace demands for emotions amidst a proliferation of compassionate service sector business models. Or, Arlie Russell Hochschild also sees the expansion of the 'new world of work' and demands for emotional labour as blurring and confusing divisions between private and occupational life in ways that that favour business interests over those of individual citizens.[11] For Fraser, who draws explicitly on Boltanski and Chiapello's thesis, this 'new world of work' grants even explicitly emancipatory discourses such as feminism a second, 'shadow life':

> At one end, [are] the female cadres of the professional middle classes, determined to crack the glass ceiling; at the other end, [are] the female temps, part-timers, low-wage service employees, domestics, sex workers, migrants, EPZ workers and microcredit borrowers, seeking not only income and material security, but also dignity, self-betterment and liberation from traditional authority. [These have been] harnessed to the engine of capitalist accumulation [such that the new left and in particular for Fraser feminist] critique of the family wage ... serves today to intensify capitalism's valorization of waged labour.[12]

In short, the normalizing of modernism, especially as expressed in the tendency to 'reject ... all forms of disciplinary regulation' in favour of personal autonomy, was extended to the support structures of the capitalist enterprise and the postindustrial state itself.[13]

In conditions of postindustrialization, a highly flexible autonomous workforce has fewer on-costs; employee creativity cannot be quantified and therefore remunerated – it is simply an employment expectation; and independent contractors do not require mandated benefits but are, rather, regarded as 'store associates' or 'overseas partners'. That is, postindustrialization and 'neoliberal' reformism combine with discourses of self-realization, creativity and authenticity to effectively de-centre the state's capacity to govern the status of the compromise on the social bond between collective obligation and individual autonomy in favour of progressive distributionary policy. In this situation, as Bauman notes, 'the dismantling of the welfare state "as we know it"' was publicly represented as 'an issue "beyond left and right", as once upon a time the creation of the welfare state used to be'. Meanwhile, the neoliberalization of the politics of redistribution has meant that the prospect of joining a poor and marginalized underclass has come to represent the sole alternative to the exhausting, emotionally draining and precarious job of 'staying in the game'. Citizens are subjectively freed to pursue almost unbounded self-sovereignty yet are bound to near-incessant maintenance of individual viability in the 'new world of work'. Similarly, Bauman argues that a new world of consumerism opens up alongside the new world of work, in which individuals called upon to keep up appearances and to be innovative and creative must consume with fine judgement, lest they fail to exhibit the 'cultural capital' that flexible and 'creativity-driven' labour markets require.[14]

Moreover, the producers of commoditized goods and services in general recognize that mass-produced goods and services are no longer as attractive as those emphasizing quality, artisanal origins and 'organic' origins in non-industrialized nature. Alongside Bauman, numerous researchers demonstrate the shift from emphases upon efficiency, calculability and control to emphases upon high-quality 'personalized' service and how the obfuscating of efficiency and calculability in favour of 'naturalness' and authenticity in consumer goods and services creates pressure upon persons to exercise a unique 'individuality'. Contemporary social conditions are such that, for Andrew Ross, atomized citizens are free to follow paths that lead everywhere, but to a stable career, free to change discipline, job title, clothing, and personality, and free to sleep under a bridge.[15]

For Boltanski and Chiapello, a 'networked' or 'projective' city or polity has emerged that, governed by a connexionist model of justice, satisfies demands emanating from within the artistic critique of capitalism, such as for creativity and mobility and respect for inclusiveness and heterogeneity, while also relegating other political concerns, including those for distributive equality and fairness, to the margins.[16] While I recognize the centrality of the 'projective network' to social conditions in the 1990s and 2000s, I later question the importance that Boltanski and Chiapello attribute to the idea of a 'connexionist' test for the assessment of justice claims by the state. Putting such reservations aside for now, it is Honneth who draws together Boltanski and Chiapello's general theoretical claim in the context of those made by the sociologists of work and consumption:

> [T]he individualism of self-realization has since [the 1970s] been transmuted – having become an instrument of economic development, spreading standardization and turning lives into fiction – into an emotionally fossilized set of demands under whose consequences individuals today seem more likely to suffer than to prosper.[17]

While calls for less conformity and greater creativity were incorporated into deregulated employment markets, with the result that work in the postindustrial 'flexible economy... requires total commitment',[18] consumer demands for authenticity have become central to an explosion of highly differentiated personal-use consumer goods and services aiming to ground the quest for self-realization in the uniqueness of subjective experience. For those concerned with the experiential and existential dimensions of the 'new worlds' of work and consumption, such as Sennett, Anthony Elliott and Charles Lemert or Gerard Delanty, for whom 'the problem of choice is now solved increasingly by the individual, whose capacity to act is coming to rest more and more on a reflexive relationship between experience and cultural options' and less on collectively ordained knowledge.[19] Relatively well-educated and articulate individuals are increasingly being asked to autonomously and creatively self-orient without referring to collective expectations, beyond those that compel one to be more one's self. In this situation, it has been argued that 'identity is in the process of being redefined as pure self-reflexive capacity or self-awareness', what Taylor sees as the search for authenticity through 'self-discovery'.[20]

In this situation, personal autonomy is no longer affirmed through social solidarity but through self-orientation. With the normalizing of modernist culture, individuals are called upon to judge the value of collective representations in relation to subjective aspirations themselves, such that it is the perspective of optimal self-affirmation that emerges as the guide for individual choices to act on political problems. This undermines possibilities for creating political allegiances that refer to class- or status-group norms or solidarity based in shared experiences of exploitation. Political ideological representations that promote collective obligation no longer gain influence by appealing directly to socio-economic classes engaged in debate over policy for redistributing wealth. Where 'a contemporary individual can only be more him or her self', appeals to the collective interest that demand class or status group allegiances become highly problematic.[21] Gathering support for redistributive rights and duties becomes increasingly difficult, because it no longer makes sense to 'orient one's self according to the perceived collective viewpoint' of a right to share in the benefits of industrialization.[22] That is, the acceptance of pluralism itself becomes, with the displacement of conformity, the basis for collective cohesion.[23] Ethico-moral ideals of the good life and how to treat others become grounded in the subject-centred notions of self-realization 'amidst the maelstrom'. At issue here is that, individualistic demands that personal self-development be met on one's own terms are most readily satisfied in deregulated employment and highly variegated consumer markets.

A key consequence of the normalizing of holistic modernist culture in the context of postindustrialization has been a truncating of possibilities for enunciating *both* modern and modernist progressive positions. The politics of postindustrialization all but exclude questioning redistributive arrangements in ways that evoke a presumption of conflict of interest between social classes as 'winners and losers'. Even though questions of redistributive justice remain central to the concerns of many progressive social movements, they no longer gain purchase in a political sense when couched in such terms. This is the situation that Fraser describes as the 'postsocialist condition'.[24] Meanwhile, the progressive political project that aimed to universalize the capacity to embrace self-realization, authenticity, creativity and the social and subjective benefits of understanding nature and culture as being reproduced symbiotically within the ecosphere has been rendered problematic. Claims for a right to self-realization have been partly addressed by the promotion of individualistic instrumentalism in the workplace. Calls for greater authenticity in everyday life are partly addressed by the expansion of opportunities to access personal-use commodities and 'entertainment'. Claims for unity with nature have been addressed by 'caring, sharing' capitalism. All the while, unjust processes of unsustainable development continue to be reproduced. As the sociologists of work and consumption show, many questions relating to justice are unsatisfactorily answered by the purveyors of new-wave management jargon and personal-use consumer goods. That is, just as many progressive political achievements within the frame of a dualist cultural ideology became ossified with the establishment of the institutional arrangements of social citizenship, it is necessary to recognize also that aspects of the progressive political programme, as articulated from within a holistic cultural frame, also appear problematic when considered in relation to the postindustrial state.

4.2 Subpolitics and risk: from 'government' to 'governance'

As van Steenbergen and Turner observed, a key impact of new social and countercultural movement influence upon the greening of citizenship in the 1990s was the drawing into political contention of hitherto elided questions of society's participation in nature. A central concept of interest in the efforts to theorize this greening of citizenship is that developed by Beck to explain the political culture of 'risk society'. Indeed, discussing the concept of 'risk' in terms of citizenship provides an important purview from which to interpret the ideological grammar of injustice in postindustrial conditions. As noted, for van Steenbergen and Turner, as well as Beck himself, as for Isin and others such as Arthur P. Mol and Gert Spaargaren, increased awareness of risk in postindustrial societies cannot be separated from the deepened or heightened individualization that accompanies societal postindustrialization. In this view, risks that are imperceptible to the senses – such as dioxin poisoning, which does not cause itching or swelling, has no smell or taste – require increased trust in the very institutions that citizens have been by and large abandoning in favour of action at the 'subpolitical' scale of individuals and local communities and which build upon inter-relational bonds of trust.[25] This is because, for Beck, a generalized awareness had emerged that the modern world increases the difference between calculable risks or hazards and non-calculable uncertainties or risks that are created as society is reproduced. For him, a consequence of widespread awareness of risk in the context of individualization is the emergence of a form of 'subpolitics that is at once "egoistic" and "altruistic"'.[26] This subpolitics is said to be 'based on the defence of life as a personal project' and the 'rejection of its adversaries; a powerful market system on the one hand and a [conformist] communalism that imposes purity and homogeneity on the other'.[27] Indeed, Beck describes a subpolitical arena wherein debate is motivated by calls for government from below and is premised upon the rejection of institutionalized politics and demands for participation and self-organization as a means of minimizing the creation of new risks and managing existing ones. Indeed, for Beck, localization and self-help, wellbeing and green political values are central to the new subpolitical ethics:

> [where] the idea of quality of life... control over a person's 'own time' is valued more highly than income or more career success. [P]roviding there are basic securities, lack of waged work means time affluence. Time is the key which opens the door to the treasures promised by the age of self-determined life; dialogue friendship, being on one's own, compassion, fun, subpolitical commitment... rest, leisure, self-determined commitments and forms of working, relationships, family life... these are the values of a self-oriented culture which is sensitive to ecological concerns.[28]

Put differently, subpolitics is the form taken by politics in the context of a modernist culture, where it no longer makes sense to situate or orient the self according to an unquestioned collective viewpoint, where what defines ethics or the good life and morality or the treatment of others concurs with ideals of seeking self-realization and authenticity amidst the disruptions of societal modernization. That is, subpolitical ideals express modernist concerns and point to the establishment of a holistic cultural ideology. Insofar as Beck's 'age of self-determined life' displaces questions of redistribution, it also displaces a shared conception of the common good as enunciated in dualist terms of domination over others and nature. Indeed, for Dumont and Gauchet, the emergence of concerns with localization and self-help, wellbeing and green political values signals an opening-up of the social question to encompass 'the socialization of life as a whole, the vital order' at the expense of dualist questions of redistribution between social classes and anchored by a culture of domination over others and nature.[29] In short, the entry into the grammar of contemporary culture of holistic discourses of society's participation in nature coincided with the dissolution of dualist 'us and them' politics. That is, Dumont's protest against the 'disruption of natural equilibriums' that placed a check on modern artificialism and Gauchet's 'remarkable return' of nature within culture coincide with what Beck sees as the challenges of a quite unprecedented scale that arise as the deep individualization that is central to subpolitics erodes the political consensus of the welfare state.

For Foucauldian theorists such as Nikolas Rose, it is indeed widespread awareness of 'risk' that has brought concerns with promoting wellbeing, community cohesion, trust and security into contemporary politics. However, for Rose and other governmentality theorists, this transformation arises not so much as a product of the pressure exerted by new social movements in the context of cultural ideological transformation but as the product of neoliberal policy strategies that support individual self-responsibility or *responsibilization*.[30] In Rose's view, such 'governmentality' is achieved because, historically, the modern state can be observed as increasingly outsourcing the administration of 'biopolitics' to individuals and communities. Indeed, Rose links the duties of the self-responsible citizen to the self-disciplining force of a biopolitics that isolates individuals and communities as subjects in relation to such all-pervasive forces of domination. In this perspective, the trans-historical force of biopolitics fosters ever-increasing 'capillary' domination, such that what I regard as modernist cultural sensibilities that privilege self-realization and authenticity emerge as the means by which subjects are bound over to develop the self-disciplining agency necessary for minimizing risks to the self, the community and the nation. What is interesting about the post-Foucauldian analysis of risk and self-responsibilization is that the theorists see evidence of the promotion of such governmentalism in demands for the localization of politics and for increased participation in governance as well as in the workings of the state.[31] This is rather strongly interpreted through the Foucauldian lens by Michael Gunder as creating a situation wherein a definition of 'sustainability', for example, emerges 'as a materialization of dominant institutional ideologies supportive of growth and capital accumulation that maintain the existing *status quo* of class inequalities, with limited regard to the environment'.[32] Or, indeed, where Mike Raco claims that governance or 'governance for sustainability' has become the vehicle for extending the 'relational citizenship identified by Rose, in which rights and responsibilities become conditional upon what an individual citizen can provide for the wider public good as opposed to universal citizenship entitlements based solely on the criteria of nationality'.[33] In short, the governmentality approaches leave little room for optimism. Lacking an explicit normative orientation, governmentality theory recognizes only the exercise of raw power.

Meanwhile, post-Marxist Gramscian analyses of the localization and community-orienting of politics around discourses of governance take a slightly different tack. Such approaches tie institutional arrangements that emphasize community and individual self-responsibility for quality-of-life and security to a particular class fraction, whose increased awareness of risk and calls for local community control are readily embraced by the advocates of neoliberalism and the 'new public management'. Adopting a line of argument that in some respects resembles that put forward by Boltanski and Chiapello, critical political geographers such as Bob Jessop, Neil Brenner and Roger Keil[34] identify historical links between demands for localization, and especially communitarianism, and the sensibilities of an urban 'creative class' that seeks to exercise a 'stake' in urban planning. This is said to take place at the cost of the poor and inarticulate, the interests of whom are marginalized from 'gentrified' stakeholder communities. The post-Marxist approach tends to link calls for local community control and demands for wellbeing with the 'post-materialistic' sensitivities of an urban elite class fraction, one that is disparaged for narrowing the contents of justice.

That is, these approaches to the questions that Beck's notions of 'subpolitics' and 'risk' raise tend to emphasize the margins at the expense of a generalizable view of the new citizenship and state form that arose in the 1990s and 2000s. This said, these approaches point to a situation in which the types of social and political participation that coincide with new citizenship claims may in fact draw self-responsible citizens together as local communities of trust that, through largely subpolitical mechanisms of governance, do facilitate quality-of-life concerns over wider considerations. Rather than focusing on the ever more complex workings of biopolitics or the predilections of a gentrifying 'creative class', it is possible to explain this shift in terms of a shared cultural ideology, one which anchors the compromise over the social bond – and, so, shared values of freedom, equality and autonomy – in holistic concerns for wellness, as security from risk, and that may but does not necessarily encompass concerns with ecospace. That is, political concerns are being encompassed within a historically ambiguous holistic grammar: key political questions become ones relating to quality-of-life, security from risk and allied notions of bonds of trust, local communities, liveability and, of course, sustainability.

Looking more closely at the widely researched shift from government to governance as a central feature of policy discourses about the greening of the state provides a first step towards seeing things in such terms. In contrast to 'government', which consists of rule-based administration that emanates from a centralized state control apparatus, 'governance' entails both formal and informal regimes. These are based on interaction, partnership and cooperation across governmental agencies, non-governmental groups and private actors, combined with policy support for self-regulation in relation to market actors based around mutually accepted voluntary compliance codes or 'opt-in' schemes. In this view, 'top-down' government, once characterized by confrontation, forcefulness and regulation has morphed into horizontal governance, which privileges reflexive deliberation, networked management and stakeholder negotiations over how policy is to be defined and implemented.[35] Indeed, governance that privileges collaboration across interest groups, dialogue-building and citizen participation in policymaking and planning has in this sense 'officially' displaced strong regulatory 'government' since the early 1990s. For theorists working on the problems posed by the 'post-political condition', the shift to governance is said to diminish, in sum total, possibilities for political action. It is in this sense that Swyngedouw argues that governance regimes are actually symptomatic of the post-political condition, 'forestall[ing] the articulation of divergent, conflicting and alternative trajectories of future socioenvironmental possibilities ... [and] solidifying a liberal capitalist order for which there seems no alternative ... [and] which eludes conflict and evacuates the political field'. In Swyngedouw's perspective, governance does not merely comprehensively displace the state's capacity to govern but displaces politics itself. For him, governance is in fact a form of 'populist fantasy', such that 'the architecture of populist governing takes the form of stakeholder participation and forms of participatory governance that operate beyond the state and permit a form of self-management, self-organization and controlled self-disciplining under the aegis of a nondisputed capitalist order'.[36] However, Swyngedouw's call for a 'radical socioenvironmental political program' seems to be at odds with the total 'forestalling' of alternatives and 'nondisputed' character of the conditions that he otherwise identifies with governance. This said, in light of his work, it appears that governance needs to be understood as at least making it difficult to extend citizen participation beyond practices that seek to increase individual self-responsibility, demonstrate but fail to normalize activist *niche* practices, or support oppressive forms of 'community securitization'.[37]

As such, I contend that the turn to governance needs to be understood in broader terms, in relation to more or less functional state imperatives and as a consequence of the cultural 'turn' that Beck sees as the source of subpolitical demands for wellbeing and security from risk. Governance in this view favours types of social and political participation in which self-responsible citizens act locally and independently to increase individual and community quality-of-life and, so, minimize or manage security from risk. As both the governmentality and the post-Marxist approaches demonstrate, governance most often involves privatized and localized social services targeting communities and individuals using various models of inclusion and participation aimed at promoting independence and trust while minimizing the consequences of risk. Governance does, however, also usher in less hierarchical, inclusive decision-making on planning, localized collaboration between actors holding different interests and consensus-building across groups that would otherwise be constrained within operational 'silos'. Governance does not preclude challenging, even if at times providing active support for, neoliberal policy, such as privatizing social services. Governance forms an integral dimension of institutional responses to citizen assertions that injustice is present in the context of postindustrial conditions. Amidst a subject-centred ethico-morality and what Beck and the citizenship theorists, such as van Steenbergen, Turner and Isin, regard as individualization and widespread awareness of risk, governance represents an institutionalizing arrangement of subpolitical demands. Governance is a response to culturally grounded forms of political opposition – such as consumer boycotts, community group pressure on city planners or demands for dialogue on the local environmental impacts of actions by global corporations – to the predominance over the 20th century of the centralized 'top-down' technocratic policymaking and rationalist bureaucratic planning that characterized the welfare state institutional arrangements of social citizenship.

That is, the means of implementing *governance* itself emerge as the key institutional arrangements of contemporary citizenship. The establishment of localized, collaborative and participatory governance can in these respects be understood as having emerged in the context of what Maarten Hajer describes as the institutional void of postindustrial societies. Where so-called subpolitics normalizes situations in which citizens eschew mass politics and bureaucratic planning in favour of local and community management and self-responsibility, the actual implementation of policy to ecologically modernize is, for better or worse, partially taking place beyond the political sphere of the state. For Hajer, this 'institutional void ... erode[s] the self-evidence' of modern institutions as the locus of politics, such that policy and practice are 'made without the backing of a polity' in the 'traditional' sense of mass-party support for nation-building projects based in 'objective' measurement, predictability and control of human and natural resources.[38] Governance has become established in response to citizen assertions that low quality-of-life and high exposure to risk, especially but not exclusively in relation to environmental issues such as pollution or climate change, are sources of injustice. And, moreover, that these issues are legitimately addressed at the subpolitical scale by individuals and local communities who recognize non-contractual relations of trust between themselves, as independent holders of a 'stake' in society. In this view, the widely debated shift from 'government' to 'governance' represents the de-centring of the legitimation of authority to organize the rules for life held in common within a particular cultural context, 'rather than a diminution of states' steering capacity'.[39] For Douglas Torgerson, the shift from government to governance is, in fact, a response by the state to citizenly malcontent with the predominance of procedural, technocratic and bureaucratic, instrumental rationalism against the backdrop of societal postindustrialization.[40] In this perspective, the shift to governance represents one aspect of broader institutional responses to the normalizing of a holistic cultural ideology, wherein the preponderance of risks and low levels of wellbeing, and dependency on the administrative competency of the state, are seen as consequences of the irrationalities of advanced industrial society and, in particular, a dualist worldview. As such, for Lorraine Elliott, the institutional arrangements that are central to governance regimes create new bottlenecks *and* opportunities at all scales:

> [Governance displays] a genuine commitment to common and shared interests, ecological values and an emphasis on human rather than state security... [and] mandates the empowerment of the 'local' and those most directly affected by environmental degradation as a way to hold accountable public and private power and authority... [However, as it stands, governance] requires greater transparency, less centralization and a more equitable sharing of wealth and technology [and] new forms of political community [more generally].[41]

In this perspective, collaborative, consensus-driven and stakeholder-centred participatory governance is the product of market-centric globalization yet also represents an institutional response or adaptation to citizenly assertions that injustice is present and definable as low quality-of-life, an absence of security from risk or a lack of opportunities for self-responsible local participation in decision-making, amongst other things.

4.3 Stakeholder citizenship and the test of wellness: a new imperative of state?

In their groundbreaking work, Boltanski and Chiapello argue that a new kind of test has emerged as the consequence of the assimilation to 'capitalism' of the 'artistic critique' of industrial society. For them, the new test is anchored by the 'connexionist logic' of a 'projective polity' that incorporates aspects of the hitherto radical 1960s countercultural worldview. The view that I propose here differs from that developed by Boltanski and Chiapello, however. In contrast to their claims that the assimilation of 'the spirit of 1968' to the operational logic of the postindustrial capitalist state explains the emergence of a new 'projective' test, I suggest that the projective model of justice is more fruitfully seen as but one aspect of a broader cultural turn to holism. This turn encompasses a diverse grammar for debating political questions of equality, freedom and autonomy, one that privileges concerns with individual and community independence and self-responsibility for quality-of-life and, by extension, security from risk. That is, I find that the postindustrial state is increasingly called upon to arbitrate the compromise on the social bond by reference to a test of *wellness*. My critique is informed by reviews of their work on 'the new spirit of capitalism' by Peter Wagner, Turner and Thomas Kemple. These critics argue that Boltanski and Chiapello tend to overlook the fact that not all citizens are always engaged in disputes over justice, that issues such as ethnicity- and gender-based discrimination are overlooked in constructing the narrow category of the projective-connexionist polity and that politics cannot be accounted for by focusing solely on networking in the occupational domain.[42]

With these criticisms in mind, when the projective-connexionist polity is seen to provide a prevailing cultural ideological frame, analysis omits consideration of situations and actions that fall outside of but impact upon politics, such as the subpolitical concerns with life-politics and risk or the rise of environmentalism or concerns with quality-of-life. Similarly, while the casualization and flexibilization of work that I discuss in Chapter 3 is a highly important feature of postindustrial social conditions, it is precisely this casualization of income and working hours that makes 'lifestyle' issues that fall outside of work so much more important, while also making occupations themselves 24–7 pre-occupations. The normalizing of modernism grounds ideological representations in shared presumptions about the importance of work in all its forms as well as leisure, play and interconnectedness with others.[43] What has shifted is a dualist ideological emphasis on maintaining and reinforcing the coherence of the self against modernization, which has given way to an ideological grammar of holistic representations portraying self-realization 'amidst the maelstrom'. While Boltanski and Chiapello account for the normalizing of pursuing an income in the service economy over drawing wages from full-time industrial employment, they do not draw out adequately the broader implications of postindustrialization. Such a view also elides the presence of increased concerns for quality-of-life, security from risk and self-responsibility that are outside of or inform the occupational domain. It is only by conceiving of the prevailing cultural ideology of the postindustrial state in these terms that demands for subpolitics, as expressed in local community-driven demands for quality-of-life and local governance, for example, become visible. The metaphor of the 'projective city' makes it difficult to recognize the impacts of ex-occupational life and the transformation of occupational life itself upon politics and culture as consequences of postindustrialization.

Other critical pragmatic theorists such as Thévenot, Lafaye and Moody argue that a green model of justice has emerged at the centre of Western culture and politics, or at least in France and the United States.[44] In the 'green' model of justice, high status is accorded to valuing wilderness and sensitivity to the limitations that the environment places upon society. Such a view concurs with arguments made in the late 1990s that the state management of the legitimation imperative has come to be defined in terms of supplying environmental conservation. The problem with locating the ideals of the green model of justice as the defining feature of the postindustrial state (a *green* spirit of capitalism?) is that such a move fails to account for reactionary arguments for the greening of the economy, as put forward by Roger Scruton, for example.[45] Conservatism, taken in this sense, wants to have its cake and eat it too, both to justify politics based in the preservation of existing privileges and to conserve resources for future use by those who will be privileged to act in unbounded ways, by virtue of desert defined in terms of effort, ethnicity, gender or religion, for example. Rather, I see the proliferation of networked relations and the ascent of green discourses emerging together, as different aspects of a broader shift towards cultural ideological holism. Rather than viewing political debate in terms of a connexionist logic that attributes status to those who can best link together nodes in a network or a green logic that prioritizes wilderness and 'limits', I regard political ideological positions as being encompassed by a holistic cultural ideology that is re-defining shared concerns with individual freedom, equality and autonomy. This perspective draws attention to 'the whole social question', and the emergence of a new compromise on the social bond that is oriented by a test of *wellness*.

This said, a key problem for critique that emerges with the establishment of cultural holism is that a test of wellness obscures possibilities for political accusations that injustices are a consequence of the specific actions taken by particular groups or classes in society. That is, in contrast to tests grounded in a dualistic ideological frame, in a holistic frame there appears to be no class or social fraction that can be held directly accountable for injustices. For example, coal-using utility companies might be responsible for pollution, but solving the sustainable development problem requires that they change. As a class, 'power generators' cannot be done away with in the same manner that many once wanted, justifiably or not, to eradicate the class 'employers'. Similarly, employers that pay low wages might be responsible for exploitation, but solving the problems of wellbeing, 'liveability' and, indeed, social participation, inclusion or mobility[46] requires more than economic redistribution from one class to another, based in claims about fairness or desert. While employers bear a responsibility for employees as a class, and coal-using utility companies for carbon emissions, the class 'employers' or 'power generators' cannot be held directly or solely responsible for unsustainable development: exclusion, low mobility and 'lifestyle' problems, climate change and resource over-use are whole-of-society problems and are the by-products of unsustainable development.

The policies, processes and practices that are associated with networked governance in this view merely reflect a broader emergent trend towards a flattening and relativizing of citizenship itself. Where horizontal national and international governance networks shape localized community-stakeholder negotiations, and where political parties appeal to citizens as the holders of 'stakes' in a globally competitive greening state that is unified around local communities of trust, it is very difficult (but not impossible) to talk politically about injustice. Hence, in making this point, I am not arguing that some kind of depoliticizing narrative of declension is at play. Rather, in contemporary political discourse, no one class or group can be accused of benefiting directly from the object at stake: for example, low quality-of-life, lack of community trust, insecurity in the face of risk, lack of individual independence or, indeed, unsustainable development. No one class or group in society benefits from these key political concerns when they are expressed in a holistic cultural frame: unlike social conformity or mass-produced 'inauthentic' goods and services or a lack of jobs and schools, low wages and unsafe workplaces, low quality-of-life, high levels of risk and low community trust and participation are not the responsibility of any one class or group within society.[47] These benefit no class or group in the same ways that 'the bosses' did benefit from the exploitation of 'the workers' in the context of social citizenship. While participation in labour unions and the 'suburban dream' were the preserve of the very strata that gave social citizenship its shape and form, and literate, landowning, adult males were that group which gave bourgeois citizenship its shape and form, the one-world vocabulary that is closely associated with the new citizenship and especially its rights and duties and institutional arrangements point to society as a whole and are the responsibility of those holding a 'stake' in it. There exists no 'class' of persons whose collective suffering from *un*-wellness is in recognizable proportion to the benefits of identifiable others. Even though the poor are harmed disproportionately more, an unsustainable society or, indeed, an 'unliveable' city poses 'risks' for the rich and the poor. Moreover, in the postindustrial ecomodernizing state, profits may accrue to polluting or labour-exploiting companies and their shareholders, but these same companies and shareholders are more than likely exemplars of 'corporate environmental and social responsibility'.

For another theorist concerned with the political implications of the networked society, Manuel Castells, 'Networks dissolve centers, they disorganize hierarchy, and make materially impossible the exercise of hierarchical power without processing instructions in the network, according to the network's morphological rules.'[48] Networked governance emerges as a consequence of the partial dissolution of 'traditional' national- and class-based power structures within and between states and favours atomistic and relativized power relations, while privileging the relatively smooth, horizontal flow across space and time of global, de-territorialized relations, especially market economic relations. However, reference simply to proliferating networked governance regimes does not clearly explain how the normalizing of collaboration across interest groups, consensus-driven negotiation and 'horizontal' deliberation amongst localized communities and individuals might be transforming politics. On the one hand, stakeholder-centred network models of governance that are premised upon collaboration, negotiation and community consensus and trust-building can fail to deal with the fact that 'consensus is not an end point nor is it always desirable', and evidence that a 'belief in win-win and...lack of attention to losers has [in many cases] undermined human welfare' and environmental outcomes.[49] On the other hand, governance is 'an open textured concept but, in some contexts, this is an advantage rather than a handicap', because it serves as a reminder that 'there may be structural flaws of social and political organization that need to be remedied'.[50] In this light, an important feature of the 'open texture' of networked governance is that it fails to define the boundary between a strong network or good governance procedure and achieving a political end, such as strong sustainable development. Collaboration, consensus-driven negotiations and stakeholder participation can sustain networks and, therefore, instrumental actions in a number of forms. For Dryzek, networked governance does not have to mean *democratic* networked governance. For him, problem-solving is accompanied by negotiation of rules and norms to govern the interactions and responsibilities of stakeholders within a debate; however, '[d]ispersed decision-making, venue-hopping...and fluid problem definitions are pervasive'. Networked governance of itself offers no grounds for assessing the quality of collaboration, consensus, participation or negotiation. As Dryzek notes, where governance prevails, good networking can merely benefit collaborating stakeholders, while it is hard to hold those who wield power accountable for their actions or for implementing any decisions reached.[51] Good networking – that is, collaborative involvement in a process – may even displace action to achieve common goods, such as when 'actual' action on environmental change is displaced by policy pronouncements. Collaboration and consensus can occlude the presence of instrumentalism that is not geared towards achieving a common social good, be it defined in narrow terms of consumer lifestyles or 'the world's most liveable cities' or broad terms of sustainable development. Seeing things in this way offers some insight into why, as some argue, the issue of 'conflict of interest' practically disappears in the prevalence of networked modes of governance.[52]

Conversely, governance also raises possibilities for categorizing political action in relation to shared cultural concerns. The institutional arrangements of governance are in this sense always *for* something that might best be defined as the provision of 'stakes' in society, such that autonomous individuals and communities are positioned to take advantage of opportunities to improve their quality-of-life and security from risk. That is, governance is in the broadest sense about 'wellness'. On this view, networked governance offers a basis for a kind of culturally acceptable 'evidence' by which assertions that injustice is present can be measured and acted upon politically, that is, tested, as contributions to strong sustainable development. In the postindustrial ecomodernizing state, networked governance efforts – such as those convened under 'Local' Agenda 21, by the International Council for Local Environmental Initiatives or by corporations practicing the 'triple bottom-line' – are by and large *for* the social good of wellness. In order to be legitimately seen as a contribution to achieving such a good, for example, actions must have consequences for achieving it, and not merely for the instrumental goal of networking or partaking in the governance process. Unjust actions are in this sense shallow, and risk being exposed as having few or low consequences as a contribution to achieving clearly defined ends. Hence, forms of governance that are justifiable have high consequences for achieving a shared goal. To be perceived as justified, governance in practice must go beyond mere acts of stakeholder collaboration, negotiation and participation. Just actions cannot be limited to simple instances of networking, which is the case in relation to many voluntary compliance schemes that offer harmless 'branding' rather than substantively binding conditions for assuming membership, or 'box-ticking' self-assessment rather than mandatory goals. Governance practices always risk accusations of shallowness, such as merely providing opportunities for networking within an NGO-industrial complex or advertising campaigns.[53] This said, there are challenges for political actors that arise when the context of the test privileges a networked structure. An individual or organization that calls directly for consequential outcomes can easily be disaffiliated as 'too radical'. Just as 'decentred, networked governance does not have to be democratic',[54] collaborative and participatory governance can be reflexive, yet not necessarily critical or democratic. Outright claims of malfeasance against collaborators may see the accuser branded as an extremist, as uncooperative or as an alarmist. Similarly, direct action such as self-help collectivization – as opposed to symbolic 'public relations' or 'agitprop' gestures – places activists in a niche situation that is outside of the network. The predominance of networked modes of governance in this view prevents a 'return' to out-and-out criticism that would identify and denounce a polluter or exploiter. Such denunciations would undermine the network that stakeholders collaborating in a governance process share. This said, some stakeholders have at their disposal the means to partially defray obligations in ways that do not undermine the network – such as when 'good corporate citizens' or supporters of 'cross-party consensus on environmental change'[55] appeal to holistic ideals of a shared world – while remaining wedded to instrumental goals.

The shift to collaboration and cooperation across interest groups within governance networks can also potentially foster open-ended deliberative governance, in which questions of power and value are open to challenge in terms of their consequences for achieving shared social goods. It is this order of practice that provides the democratic groundwork for tightening up and, so, categorizing action in relation to the test that I am associating with the types, contents and arrangements of contemporary citizenship. In this view, if the state is to continue as the legitimate authority that organizes the rules for life held in common in the context of the new citizenship, it is bound to provide conditions that support culturally meaningful notions of freedom, equality and autonomy that are grounded in a 'test' of wellness which encompasses holistic concerns with individual self-responsibility in relation to questions of quality-of-life and security from risk. The types of social participation, the contents of the rights and duties and the arrangements of contemporary citizenship can in these terms be interpreted as consequences of contemporary institutional responses to what Wagner delineates as three separate but interwoven dimensions of the legitimation crisis that was observed beginning in the 1970s. These are the 'crisis of political accountability', the cultural 'crisis of meaning' and the 'economic crisis of the satisfaction of needs'.[56] Seen as an institutional response to the political crisis of accountability, the new citizenship implies a transformation from top-down abstract rule-based administration to the embrace of local participation and community demands for the means to appraise and address global 'risks' on the basis of trust and civil law non-contractualism. Seen as a response to the cultural crisis of meaning, what I define as 'stakeholder' citizenship implies the primacy of authentic self-realization in relation to, and privatized self-responsibility for, environmental and social problems, over an emphasis on respect for objective bureaucratic-rational procedure and conformity with some 'public' ideal of collective cohesion. That is, the new citizenship privileges ethico-moral obligations and a conception of the postindustrial ecomodernizing state in primarily voluntarist terms, as an ethico-moral community that regards itself as political. And, seen as a response to the economic crisis of the satisfaction of needs, contemporary citizenship implies the primacy of self-regulatory voluntarist self-responsibility as the prevailing means for satisfying needs through 'deregulated' global markets, displacing the satisfaction of needs through stability and top-down regulation-based redistribution from those who benefit most to those who benefit least from market involvement, as defined by state boundaries. That is, while in the early 1990s it appeared that 'green' concerns with natural heritage or attentiveness to environmental conservation[57] might have emerged as the defining feature of the legitimation imperative of the state, by the 2000s it seems that the situation had changed, and the legitimation imperative has since required the state to foster opportunities for individuals and communities to take advantage of the 'stakes' that they hold in society, primarily by providing opportunities for self-responsibility in relation to questions of quality-of-life or security from risk.

Indeed, institutional efforts to ground shared conceptions of the common good in the greening of economic growth respond to widespread acceptance by the state that its citizens regard as *virtuous*[58] issues such as 'good' governance, independence, trust, liveability, participation and sustainability. These virtues are at best imprecise – most notably, 'sustainability' offers little in the way of directives about the actions that should be carried out to achieve it – while pursuit of them implies a shared cultural holism that supports representations of society in terms of one world participating in nature. While 'risk issues can threaten the legitimacy of the state and the political economy in which it is embedded' and, so, prompt legitimation crises,[59] risks need also to be seen as having emerged as public concerns against the backdrop of a wider cultural transformation that supports subpolitical demands for quality-of-life and for views that 'there is more to life than a job for life', for example, as well as for local participation in community-building based on trust rather than bureaucratic procedure. Moreover, what cannot be ignored is that increased awareness of wellness issues, such as security from risk, and demands for individual and local independence also accompany massively expanded markets for personal-use consumer goods and services. Put differently, assertions that injustice is present gain purchase upon contemporary institutions as whole-of-society claims about well-being or liveability, participation, inclusion or trust, and tend to focus on the impacts of global risks upon local and community quality-of-life. The process of satisfying the legitimation imperative in the postindustrial ecomodernizing state thus differs from the 'traditional' legitimation imperative. Whereas the legitimation crisis of the 1970s was characterized by the withholding of labour or votes – citizens acting to assert that reduced state protection of social or individual welfare is unjust – by the 2010s it has come to be characterized by subpolitical 'withdrawal' or what Gauchet describes as 'civic desertion' from politics. Stakeholder citizenship provides a basis for citizens to assert that injustice lies in state failure to provide opportunities to individuals to take advantage of the 'stakes' that they hold in society (see Table 4.1).

Table 4.1 Imperatives of state, after Dryzek and Dunleavy

Period	**Key functions (cumulative)**
Mercantile state	*External security*: Compete with other states *Internal order*: Prevent civil conflict *Raise revenue*: Levy taxation necessary for maintaining external and internal order
Industrial state	*Accumulation*: Support capitalistic market expansion *Legitimation*: Provide social welfare and,
Postindustrial ecomodernizing 'competition' state	*Provide* individual opportunities to take advantage of 'stakes' held in society

Source: After J.S. Dryzek and P. Dunleavy, *Theories of the Democratic State* (New York: Palgrave Macmillan, 2009).

As set out in Part I, in the context of mercantile capitalism and bourgeois citizenship, the parameters of the 19th-century political debate were defined largely in terms of a dualist cultural ideology that framed equality, freedom and autonomy as values that supported atomistic participation in a Social Darwinist economic struggle. High status in this sense accrues to actors making rational decisions and right ethico-moral choices, such that debates over justice coalesce around tests oriented to defining the laws of nature. Assertions that injustice is present against the backdrop of these conditions created conditions for the institutional arrangements of bourgeois citizenship: the Rule of Law, limited-representative Parliament and capitalistic market-based property rights. Similarly, as I discuss it in Chapter 3, a dualistic cultural ideological principle of redistribution had become central to politics in conditions of industrialization, in the late 19th and early 20th centuries anchoring debate in differences over the allocation of the profits of industrial capitalism between social classes. That is, where distribution is seen as a key cultural ideological concern, the high status given to freedom, equality and autonomy in the West tended to accord with access to wealth, be it the product of hard work or a 'fair share' or used to provide hospitals or schools. Legitimacy in this sense was balanced by facilitating a relatively interventionist state through the institutional arrangements of welfarism and formal contents of social citizenship. In this perspective, working-class movements sought freedom from domination in these terms and, in so doing, contributed to shaping a new form of social citizenship, one that nonetheless remained anchored by modern dualism. Social citizenship in this sense arose in the context of a dualist cultural ideology that framed debate over freedom, equality and autonomy in terms of a grammar of economic redistribution between social classes. Following this narrative of change over time, I develop the view that by the late 20th century, the impacts of new social and countercultural movements, in asserting the presence of injustice in new terms, expanded both cultural and political representations of justice to incorporate holistic concerns. That is, I describe how redistribution, as a shared value, had come to be regarded as excluding modernist cultural representations that economic wealth alone does not satisfy concerns with the good life or that recognizing minority claims and the impacts of society within the ecosphere are legitimate political topics.

In the society of stakeholder citizens, high status accrues to those actors who can be described in terms of a holistic discourse of wellness.

Low status accrues to those identifiable as *dependent* (on the state, such as are many migrants or women, for example), *unwell* (those who ignore or are ignorant of advice to eat healthy food or develop a career, for example) or *insecure* (extending from a lack of self-confidence in relation to life goals to high levels of exposure to environmental pollutants or toxins, for example). The *contents* of the new citizenship that has arisen centre upon rights to wellbeing and security from risks and duties to independently assume self-responsibility for addressing these (see Table 4.2). My point is that these rights and responsibilities may or may not include reference to justice defined in terms of ecospace. That is, while these are cautiously welcome developments, when it is seen against the backdrop of the postindustrial state, the emergence of such rights and responsibilities makes it difficult to claim that wellness extends beyond personal health or wealth, or that the provision of personal choice and local action themselves does not necessarily constitute the achievement of justice. Moreover, emphases on individual quality-of-life and self-responsibility make it difficult to differentiate clearly between discourses and actions that have as their objective privatized ends, such as profitability, or socialized ends, such as environmental reparations or social justice. That is, the contents of citizen rights and duties make it difficult to identify a class or group that is responsible for injustice. Injustice, when cast in terms of a 'test' of wellness, has no identifiably responsible class of agents or *perpetrators*, and is *diffuse* in terms of duties to minimize it. If an action or discourse is to be regarded as culturally *and* politically legitimate in a holistic frame, it must support freedom, equality and autonomy in respect of wellness. It must provide a basis for ranking status in a 'real' situation – regardless of whether the test of wellness is addressed in reactionary terms as justifying pursuit of narrow self-interest by the deserving or in progressive terms through freedom, autonomy and equality derived from fair use of ecospace, full recognition and adequate representation or participation. In part a consequence of the successes of new social and countercultural movements acting to normalize modernism in the postindustrial West, holism poses cultural ideological questions that frame political debate within a grammar that, in a historical perspective, is ambiguous. That is, the recrudescence of holism necessarily encompasses the possibility that social conditions are a consequence of society's participation in nature – what Dumont and Gauchet regard as a return to questioning 'the social whole' – but only contingently encompasses the phenomenal spectacle of the arrangement of social reproduction as the cause of social injustices or the degradation of the ecosphere itself.

Table 4.2 Marshallian significance of citizenship forms, after Susen and Turner

Citizenship form	Consolidated contents, types, arrangements and tests
Bourgeois	*Civil* dimension of citizenship guarantees legal and juridical *rights*, and emerged from challenges to the arbitrary power of absolutist regimes in the 17th and 18th centuries, alongside *duties* to obey the law and accept taxation. *Arrangements:* law courts 'all equal before the law'. Combined with *political* dimension of citizenship, consolidated formal participatory and electoral *rights* and emerged from challenges to elitist parliamentarianism in the 18th and 19th centuries alongside *duties* to select a representative and maintain basic knowledge of political affairs. *Arrangements:* parliamentary democracy elected by universal adult suffrage. *Types* of social and political participation centring on free, equal and autonomous individual who engages in the 'economic struggle' on 'legal' terms. *Test:* laws of nature.
Social	*Social* dimension of citizenship consolidates *rights* to economic welfare and material security and was established through challenges to the exploitation of labour in the late 19th and early 20th centuries, alongside *duties* to serve in conscripted armed forces, reproduce offspring and reject 'communism'. *Arrangements:* welfare system. *Types* of social and political participation centring on individual worker-consumer and 'nuclear' family member. *Test:* wealth.
Stakeholder	*Stakeholder* dimension of citizenship consolidates *rights* to quality-of-life (which includes security from risk) established through challenges to social conformity in the late 20th and early 21st centuries, alongside *duties* to independently take advantage of opportunities in relation to stakes held in society. *Arrangements:* horizontal governance networks. *Types* of social and political participation centring on self-responsible, independent citizen, acting locally through their 'stakeholder' communities. *Test:* wellness.

Sources: T.H. Marshall, *Class, Citizenship, and Social Development* (New York: Free Press, 1965); ——, *The Right to Welfare and Other Essays* (London: Heinemann, 1981); S. Susen, 'The Transformation of Citizenship in Complex Societies', *Classical Sociology* 10, no. 3 (2010): 259–85; B.S. Turner, 'Outline of a Theory of Citizenship', in *Citizenship: Critical Concepts*, ed. B.S. Turner and P. Hamilton (London: Routledge, 1994 [1990]), 199–226.

5
Not Just the Warm, Fuzzy Feeling You Get from Buying Free-Range Eggs...

5.1 The new citizenship: a 'win–win' scenario

In the 1980s and 1990s, political parties began to enunciate new discourses in their appeals to citizens. In short, across the postindustrial states, the major parties began in earnest to place modernist concerns with self-realization at the centre of their public representations and, eventually, their social policies. Indeed, where the normalizing of the subpolitics of risk helped to cement new institutional arrangements of citizenship in horizontal governance networks, what was regarded at the time as a 'new' form of Third Way politics, ostensibly steering a course 'between' Left and Right, brought with it new types of social and political participation and new 'stakeholder' citizenship rights and duties. My point is that, although differing in important ways across different states, the major parties began in the 1990s to underplay appeals to citizens as the bearers of 'liberal' civil and political rights and duties and especially 'social' rights and duties in relation to the welfare state. Rather, the major parties took up new calls to 'stakeholder' citizens as the bearers of rights to wellbeing and security from risk and duties to be independent in relation to opportunities to participate in society. Later in this chapter, I look more closely at three examples of contemporary social practice – corporate social and environmental responsibility, green consumerism and the political use of liveability indices – and find that the new citizenship is indeed premised upon a holistic cultural ideology, as well as non-contractualism, non-territorialism, the dissolution of the public-private split and ethico-moral awareness of environmental problems such as risk or, even, ecospace.

Beginning in the 1980s, with few exceptions, political parties in the settler societies and in Western and Northern Europe turned away from concerns with redistributive politics and embraced so-called Third Way ideals of enlightened self-responsibility. More recently, the central themes of the Third Way have been seamlessly continued in the emphases of Big Society policies in Britain, for example, and to some extent in attempts by President Barack Obama to maintain a 'conciliatory' presidency in the United States.[1] As is well known, the common thread linking party platforms across the West since the 1990s has been an emphasis on workforce participation, displacing that previously given to welfare provision. Other similarities appear in the extolling of the benefits of inclusion in local communities and entrepreneurialism built around reciprocal relations of 'trust' and, to a lesser extent and more unevenly across different states, the embrace of very weak or neoliberal ecological modernization.[2] Indeed, what is interesting about politicking and policymaking by the mainstream parties that have sought to 'go beyond' Left and Right since the 1990s is that citizenship itself is represented as an instrumental benefit that accrues to individuals, their communities and businesses through the self-responsible and enlightened exercise of instrumental preferences within a 'greening' competition state. The new citizenship that was established also tended to flatten or relativize the concept, insofar as all social actors – subnational institutions such as cities, major corporations or small businesses, civil and charity groups, as well as local communities and individuals in the affluent North or the impoverished South – came to be portrayed as holders of the same interests, to the same degree.[3] Indeed, its proponents define stakeholder citizenship rather ambiguously, as 'the ethical and human capital development of the self organized around the possession of stakes' in society.[4] The key premise here is that improving the economic and social capabilities of individual citizens will produce a more efficient and productive, cohesive and inclusive, socially and environmentally sustainable, globally competitive society. That is, citizens and their communities in the 1990s become the denizens of an irresistible, juggernaut-like globalizing 'stakeholder capitalism'.[5] For the major parties and the governments that they formed, the common good came to be represented as the product of well-managed markets and the provision to citizens of opportunities to participate self-responsibly in a society geared to supplying 'existential' needs, most often through activity in the economy but also through community participation, local engagement and social inclusion or cohesion.[6] The state, through such government, in this sense gestures towards but is not bound directly to address socially created injustices. Rather, state policy channels the 'trickle-down' of individual opportunities while it simultaneously manages, often with great public fanfare, the transition to a globally competitive and greening economy:

> The benefits of economic performance may not be distributed equally, but they do accrue to agents throughout the economy and have often been combined with social provision and environmental protection without this significantly damaging economic performance.
>
> [Thus i]ndividuals well endowed with economic and social capabilities will be more productive; companies that draw on the experience of all of their stakeholders will be more efficient; while social cohesion within a nation is increasingly seen as a requirement for international competitiveness.[7]

Indeed, by the 2000s, based in a perceived new politics of supplying the conditions for creating wellbeing and minimizing risk, politicians called upon 'enlightened stakeholders' to eschew welfare state institutions and to independently pursue personal goals beyond merely increasing economic wealth. These included building trust and 'getting involved' in creating liveable communities. Indeed, local community participation in the 'social' or 'green' economy provides an important discursive motif for political parties and governments addressing citizens. In the society of stakeholder citizens, one cannot be 'well' without being self-responsible. Insofar as the major parties and market actors have succeeded in representing an over-protective 'nanny state' as undermining rights to independence, one consequence has been that some degree of responsibility for securing the self from risk has emerged as a necessary condition for social participation. While the shift away from welfare provision and towards self-responsibility is well documented in the rhetoric of the major Third Way parties of 1990s,[8] for Supiot it currently combines with concerns for global competiveness amongst states:

> The model of wage labour that held sway during the industrial era – in which a worker abdicated a degree of freedom in exchange for a certain amount of security – is no longer applicable today. [T]he question [in today's 'competition state'] involves not simply the codification of the individual worker's rights but rather the creation of professional conditions for people such that, over the long term, their capabilities and economic needs are sufficiently assured to allow them to take initiatives and shoulder responsibilities. The

> key terms within this perspective are not jobs, subordination, and social security, but work (understood in all its forms, not just as wage labour), professional skills and economic security...
>
> The well-known consequence is a [global] race to the bottom in fiscal, social and environmental deregulation...[9]

That is, following Supiot, while the major parties may represent themselves publicly as entering into a 'new' social contract, they in point of fact articulate the virtues of a particular kind of civil law non-contractualism. This is especially the case in relation to contemporary Conservative parties.[10] While relativizing the status of individuals in relation to the state and market actors, citizens are drawn into the global competition between states as if they were members of a team that is, for example, 'going for green' or, as I discuss later, competing to become 'liveable'. Such representations of the new citizenship appear central to British Conservative-Liberal Democratic coalition government arguments that increased personal wealth alone, including the benefits provided by the social security system, 'cannot guarantee progress or happiness or personal fulfillment'. As such, the Big Society programme aims to 'to create a climate that empowers local people and communities, building a big society that will take power away from politicians and give it to people'. The same government in May 2011 committed the UK to cutting emissions by 50% by 2030 through a range of decarbonizing market mechanisms and green consumption programmes.[11] In the United States, in part responding to the burgeoning of the reactionary Tea Party movement, Democratic President Obama announced that 'the time for debate over the role of government and of markets in society is over', and that his administration must work to 'restore the vital trust between a people and their government'. Indeed, President Obama claimed that 'serving our country is... something that we should expect all of us as citizens to do', and that 'we are not acting as good stewards of God's Earth when our bottom line puts the size of our profits before the future of our planet'.[12]

Indeed, what this brief account of the emergence and establishment of 'stakeholder' citizenship demonstrates is a shift in the types of social and political participation supported by the state in relation to citizens since the late 1990s. Contemporary citizens are in this view not the self-interested social atoms of nation-building conservatism, nor are they the agents of social solidarity championed by progressives against the backdrop of a dualist cultural ideology anchoring politics in issues of distribution. Rather, contemporary citizens are enlightened and self-responsible individuals who – whether traditionally seen as employers, employees, offshore suppliers or welfare recipients – hold 'stakes' in communities of trust and inhabit a globally competitive greening state. Whereas social citizenship had been anchored by employment-based participation within a productive, market-oriented and nation-building state, the new citizenship that was established in the 1990s and 2000s centres upon empowered and self-responsible individuals and local community-type bonds of trust, especially in relation to independently obtaining an income and making enlightened consumption choices that build upon the possession of stakes in society.

In this sense, there exists a 'dark side' to the new citizenship, one that in large part rests in novel forms of social non-contractualism and de-territorialized global competition between greening and non-greening states. Indeed, writing in the early 1990s, Dumont proposed understanding such a situation in terms of a view that 'we live everywhere in what might be called a post-liberal regime...a limited or transcended liberalism that combines in varied proportions individualism and its contrary [holism] reintroduced, however reluctantly, by reason of necessity'.[13] Increasingly since the 1990s, citizens needing to access social services were called upon to demonstrate their willingness to act as self-responsible individuals, while, as stakeholders in a firm, individuals were increasingly called upon to deal with employers on a one-to-one basis. The centrality of this social non-contractualism to the types of social and political participation of the new citizenship is demonstrated most clearly in the 'downloading' of social services and implementing of 'workfare' regimes and the deregulation of employment markets generally that was extended across the West, beginning in the United States and Britain in the late 1980s, in Australia, New Zealand and Canada in the 1990s and in France, Germany and the European Union generally in the 2000s.[14] This is not to argue that 'welfare' is completely dissolved but, rather, that 'claims upon the state have increasingly been linked to judgements of individual behaviour'.[15]

Insofar as state commitments to maintaining welfare and employment within a nation-building state give way to social non-contractualism and the exposure of citizens to globalized employment markets, holism is in this sense 'reintroduced'. State support for modern dualist concerns with maintaining the coherence of the self 'amidst' the maelstrom of modernization – through the redistributive institutional arrangements of the welfare state – give way to state support for holistic concerns with the achievement of self-realization by capitalizing on the opportunities that holding a 'stake' in society makes available. Of course, a key concern here for the advocates of progressive politics is that some stakeholder citizens are more able than others to realize the self in local communities of trust, in obtaining an 'income' as self-responsible 'entrepreneurs of their own becoming'[16] or as 'enlightened' consumers, for example. Social non-contractualism in particular can be regarded as an aspect of recrudescent holism, insofar as state support for individual citizens, drawing upon the dualist premise of upholding formal equality between social 'atoms' *against* nature, gives way to state support for a meritocratic hierarchy of stakeholders that is grounded in a premise of competitive participation *in* the globalization process. Indeed, as I have suggested, such values are central to what van Steenbergen and Turner amongst others regard as the 'greening' of citizenship that took place as a partial consequence of the spread of new social movement actions and normalizing of countercultural values in the 1980s and 1990s. In this view, a 'fourth dimension of citizenship'[17] has indeed been established alongside those of civil, political and reduced social rights and responsibilities, but, in contrast to the aspirations once held for it, the new holism that it represents is only *potentially* commensurate with sustainable development.

In this perspective, 'stakeholder' citizenship is the form that citizenship takes in the context of postindustrial state commitments to policy discourses of very weak or neoliberal ecological modernization. Such discourses of ecological modernization are premised upon a capacity to support the proliferation of what Barry calls 'win–win' scenarios; that is, ecological modernization portrays economic growth and environmental sustainability as commensurate, without requiring structural economic change. The primary means for attaining such 'win–win' scenarios is through technological advancement, efficiency gains and a shift in consumption choices. This is what Joan Martinez-Alier refers to as the gospel of eco-efficiency.[18] In addition, ecological modernization relies upon a purely 'supply-side' policy rationale and, as Barry argues, 'does not engage with consumption issues of either challenging or regulating the demand for goods and services in the economy, or with issues concerning the distribution of consumption within society'.[19] The key point is that the political-economic structure upheld by the state and the ends to which it is oriented are, it is purported, being 'steered', however slowly or indirectly, onto a sustainable pathway without responding to calls for a radical new political-economic system. Nonetheless, as Dryzek and colleagues demonstrate, in adopting policies that are based in discourses of ecological modernization, the state does implicitly recognize calls for the 'greening' of the state.[20] It is in this sense that

Schlosberg's observations of the progressive achievements of the environmental justice and environmentalist movements that I discuss in Chapter 1 are important. The actions of these movements in achieving environmental rights and ecological guardianship can be regarded as possible and effective *because* their claims are articulated within a wider cultural ideological transformation that sustains an expanded political definition of injustice. It is in just such terms of an expanded definition of injustice that the major parties and, as I will suggest, market actors address citizens as stakeholders in a greening competition state. Indeed, drawing upon a holistic cultural ideology in their appeals to the need for non-contractual community bonds and de-territorialized global space, as well as for 'enlightened' private choices and a 'win–win' situation in relation to green demands, the major parties themselves oppose dualistic discourses of redistribution between classes engaged in conflict over the spoils of industrial production. Moreover, it is in these terms that the greening of the global competition state is legitimated as the offer of opportunities to take advantage of a 'stake' held in society. As discussed in Chapter 4, the rights and duties of the new citizenship address assertions that the risks that communities and individuals confront are unacceptable, that there is more to life than wealth and that authentic self-realization, in particular through subpolitical forms of association, is essential to social existence. To reiterate, the contents of contemporary citizenship *rights*, which centre on entitlements to individual and community wellbeing and security from risk, are complemented by *duties*, which centre on self-responsibility for independently taking opportunities to utilize stakes held in the postindustrial ecomodernizing state.

Indeed, as Barry and Peter Doran argue in relation to the situation in Britain and Schlosberg and Sara Rinfret argue in relation to the situation in the United States, discourses of very weak ecological modernization themselves have increasingly come to centre on the provision of security from risk while supporting supply-side policy to facilitate consumer choices to assume self-responsibility in relation to consumption and production/distribution practices. Economic security in this sense implies not merely the capacity to accumulate wealth by acting in markets but the assurance that markets will continue to function as the source of accumulation in the face of future risk scenarios, such as oil price rises or extreme weather events. As such, the implications for citizenship rights of the term security undergo a subtle transformation. Whereas in the context of social citizenship, economic security referred to the distribution to individuals and families of the right to maintain a standard of living, economic security in the context of stakeholder citizenship refers to a right to the benefits that accrue to individuals when, for example, energy insecurity is minimized or avoided. Especially in the United States, economic-environmental 'security is being directly linked to ecological modernization policy suggestions – in particular arguments around energy efficiency, new energy technologies, energy independence, climate change, foreign policy, and international relations'.[21] That is, citizen rights to a 'fair share' of the social product are being transformed into indirect citizen rights to opportunities to benefit from state policy aimed at minimizing the uncertainties and risks associated with social reproduction. Meanwhile, citizen duties are being recast as voluntary responsibilities to 'choose' to consume with green discrimination and adopt eco-efficient household technologies, for example, or to insure oneself against future uncertainties by participating directly in financial markets through pension plans, for example.

In this view, and in contrast to views that it is no longer possible to delineate clearly reactionary and progressive positions, I find that the new citizenship reifies the two positions. On the one hand, whereas reactionary politics had previously sought to 'conserve' or preserve existing structures of privilege by justifying engagement in politics based in claims to be upholding tradition, particularism over universalism and essentialism over contingency – and, thus, control over the world by the self – such indirectness is no longer necessary.[22] In the context of contemporary citizenship, calls to preserve existing structures of privilege refer directly to subjective freedom of choice to consume resources, for example, and social policy that will further 'unburden' individuals and communities in their narrow pursuit of self-interest, as with the privatizing of security from risk. From within this perspective, what is required are ever-expanded negative freedoms to realize unbounded selfhood. Such post-dualist conservatism is given its most visible expression by the Tea Party movement of the United States, but also in the Netherlands, Australia and Britain, for example. The Dutch politician Geert Wilders and the Cronulla Beach rioters make clear that what is at stake for them is not so much the need for strong leaders or exclusive ethnicity, as had been central to earlier iterations of reactionary politics, but the perceived threat to 'our way of life' that is allegedly posed by groups who do not embrace narrow self-interest framed by consumerism. Similarly, for the British National Party splinter organization the English Defence League, support for an ultra-libertarian interpretation of freedom places the 'right' of the English citizen to a kind of untrammelled consumer philistinism at the centre of its political programme.

On the other hand, whereas progressivism had sought to universalize the possibility of freedom by equivalizing economic redistribution and providing 'social security', on the basis of the claim that all are free, equal and autonomous citizens of a political community that grounds itself in a state that is committed to subjugating nature, the social solidarity that underpinned such claims is no longer available. From within the progressive perspective, the new citizenship implies a rethinking of the idea of employment or, indeed, of the nexus of employment and income, of access to life-long education and professional skills, as well as of support for capabilities that will facilitate independence and as such, mobility, inclusion and participation in social life generally, in all of its global interconnectedness. As such, wellness emerges as a social problem of the provision of capabilities and capacities to confront risk issues in ways that respect future generations and the ability of the ecosphere to support society at a global scale – that is, of sustainable development. Moreover, in the wake of new social and countercultural movement achievements, no longer can a single group in society – the ideal-typical white male industrial worker – monopolize the discourse of citizenship to the deliberate or accidental exclusion of others. In this view, the progressive political position is indeed bound to support political institutions that redistribute rights fairly, especially to a share of economic prosperity as defined by ecospace, recognize as equal all social participants and provide transparent representation or effective participation while delineating the territorial reach of the political community of citizens against global aspirations for sustainability. Moreover, following Fraser and Schlosberg it is not only the 'substance' but also the 'frame' of justice and injustice that is disputed, such that progressive assertions that injustice is present require domestic and international support for policy that enables capabilities for human flourishing at the subjective level and for equality of capacities to remain within the constraints of the ecosphere to support human life for present and future generations at the societal level.

5.2 Corporate social and environmental responsibility: holism and capitalism

In this and the following sections, I discuss the corporate social and environmental responsibility movement, 'green' consumerism and the political use of commercial liveability indices in order to link more closely my account of the relationship between the types of social participation, contents of the rights and duties and institutional arrangements of the new citizenship with cultural ideological holism, non-contractualism, non-territorialism, privatized political obligations and quality-of-life concerns that may but do not necessarily refer to ecospace.

Closely associated with the 'shift to governance' has been the establishment of a broad and deeply influential global corporate social and environmental responsibility movement around the ideals of 'triple-bottom-line capitalism'.[23] Indeed, many global businesses and all of the major independent global corporate social and environmental responsibility reporting initiatives – the Standards International AccountAbility Standard, the Global Reporting Initiative, the International Standards Association's ISO 26000/SR and the United Nations Global Compact – make specific reference to their signatories' embrace of stakeholder citizens, stakeholder interests and horizontal governance-type negotiations with individual stakeholders and stakeholder communities.[24] This movement taps into an important aspect of the discourses of very weak ecological modernization and the new citizenship – that is, the view that the combination of high-technology development and ever-further rounds of marketization and privatization will automatically bring society out of a largely depressing and dirty, polluted and uninspiring, socially homogeneous, heavy-industrial past: one that failed to compete in globalized markets and offered little opportunity to innovative and creative stakeholder citizens while breaking down community bonds and replacing them with 'faceless' bureaucracy. The various triple bottom-line programs that are practised by individual firms and the major global inter-firm voluntary codes of conduct, such as the United Nations Global Compact or the Global Reporting Initiative, refer explicitly to the failed dualism of industrial society's adversarial approach to dealing with social and environmental issues and advocate for 'integration'.[25] Moreover, the notions of social responsibility and corporate or business ethics promote an image of corporations as voluntary collectives whose members share the same values and aspirations and hold in common reciprocal obligations that depend not on rules and discipline enforced from the 'top down' but on consensus and shared responsibilities to deliver the best products or services.[26]

Indeed, business journalists seeking to define the 'movement' see the 'leaders of companies that shape up as socially responsible' as expressing views that 'sustainability is not just the warm, fuzzy feeling you get from buying free-range eggs':

> It's a management philosophy that ensures the continued, long-term success of a company... sustainability takes in a broader base – the

> stakeholders ... not just a company's investors but also its employees, customers, suppliers, and the community at large. Here's why. The fall of communism and the triumph of the free-market have slashed the influence of nation-states. Big business holds the reins of the world economy and, to a large extent, world politics. In an era of globalization ... information technology is bringing transparency and fluidity to global communications. Where in the past they would not have given a flying forethought to sustainable development, today's global consumers not only know what's happening across the ocean some two to 12 thousand miles away – they also care.[27]

Indeed, since the late 1990s, 'sustainability' has increasingly been understood from the perspective of 'smaller spatial scales (such as cities or regions) and economic entities (such as sectors or firms)'.[28] Meanwhile, for former Greenpeace leader Paul Gilding, market actors increasingly aimed 'to go beyond traditional single "financial" bottom-line [accounting] methods' to measure impacts of business activities 'on the social fabric, environment and, in some cases, human rights' while maintaining 'competitive advantage' in capitalistic markets using holistic '"full-cost" financial accounting methods' that 'internalize various social and environmental externalities' as business costs. For Gilding, corporations, and, indeed, 'global capitalism itself, must go beyond financial obligations to shareholders and account for their activities in support of the people who work for them, the communities in which they operate, and the global environment'.[29]

As such, and resembling the terms on which the contemporary political parties address citizens, the corporate social and environmental responsibility movement engages with shareholding owners, managers, salaried staff, waged employees, local, national and offshore contractors, and producer and consumer communities as well as individuals in holistic and non-contractual terms. The proponents of corporate social and environmental responsibility thus do not merely promote a new globalizing order based around the activities of 'caring, sharing corporations' but actively pursue a 'new era of capitalism' that aims to promote sustainability by enhancing opportunities for self-responsible individuals. The practices and discourses associated with it often intertwine with calls to 'unite in a shared purpose' and supplant a failed and substantively different antecedent. Implicit here are claims that top-down management, full-time workforces and, importantly, big bureaucratic governments restrain good corporate citizens and limit market relations to archaic, oppressive, hierarchical and conformity-driven, largely, dirty and heavy, industrial practices. This so-called new era is said to unleash a nature-like tendency for high-technology, services and consumption-centred markets to engender unprecedented opportunities for personal imaginativeness and creativity, individual freedom of choice, community development and ecological reparations and to increase democracy.[30] That is, the discourses promoted by the corporate social and environmental responsibility movement draw explicitly upon the prevailing cultural ideological holism.

Prior to the global financial crisis of 2008, the focus of triple bottom-line reporting was directed towards corporate social responsibility. In 2003, major global accounting firm KPMG explained its commitment to corporate social responsibility in these terms:

> *Understanding Corporate Social Responsibility* – The concepts of CSR and 'sustainability' are based on using and developing natural and social resources in ways and at rates that meet current needs without impairing the ability of future generations to meet their needs. Organizations focused on CSR balance their performance over three general categories: economic, social, and environmental. Thus, while CSR accounts for economical performance, it also highlights the need to respect the integrity of social systems and the rights of individuals, preserve ecosystems that support life, and conserve natural resources.[31]

After the crisis, firms such as KPMG continued to recognize the importance to creating a 'share price premium' of taking into account both social and environmental responsibility. By 2010, this recognition had come to be expressed directly in concerns with 'sustainability', which are worth quoting at length:

> The days of purely measuring business performance by financial result may well be numbered. In its place, ... discerning investors will look for something broader to measure an entity's real contribution and performance. That something could be in the shape of the 'triple bottom line'; an amalgam of financial results and an assessment of the social and environmental impacts of a business. Or, put another way: People, Planet and Profits. If that sounds like something out of a pre-credit crisis Corporate Social Responsibility (CSR) manual, that's because the concept of the triple bottom line has been around for some time and, yes, it has echoes of the CSR debate from the

> previous decade. CSR however was more an internally focused view; something for employees to feel part of and probably made up just a couple of paragraphs in the annual report. It's unlikely that investors and lenders were ever making decisions based on a company's CSR initiatives... However, the point here is that the crisis was a wake-up call, highlighting just how intertwined the fortunes of businesses, communities, resources and the man on the street have become. The crisis made investors and stakeholders far more aware of this inter-dependence than ever before; hence the growing desire to look deeper into a company's performance and beyond the numbers. Now, the good news in all this is that for those businesses that can grasp this nettle and somehow come to terms with supplying relevant multi-dimensional information to investors, I believe that a share price premium beckons...[32]

Indeed, the focus of KPMG's Climate Change & Sustainability Services 'practice is to assist enterprises in better understanding the complex and evolving business and regulatory risks relating to climate change and sustainability, while helping them to capitalize on the resultant commercial opportunities'.[33] The attention given to stakeholders in such discourses draws upon a holistic discursive frame that sees all social actors, regardless of scale or interest, as acting together voluntarily to maintain economic growth or, in the case of the firm, boost profitability, while addressing shared concerns with an ambiguously defined conception of the social good.

Of course, the triple bottom-line 'reporting' that underlies the corporate social and environmental responsibility movement is problematic for many reasons, not least because it unquestioningly sets the rules and norms associated with capitalistic markets as the benchmark when evaluating a firm's activities. The approach thus presumes a strong commensurability of values that extends across the state-economy nexus to encompass the polity. The orientation of such approaches means that strong commensurability and strong comparability of values are taken as givens. All actions and their effects are calculated under the same *numéraire*.[34] It is as if the instrumentalism of an actor carrying out a specific action did not impact upon how society affects the ecosphere. That is, as if the political and economic tensions that are created when acting to achieve a goal, such as profitability, did not produce multiple situations in which the goals of those affected were rendered incommensurable with the goal of the business actor or in which others were not affected in negative ways. Understood in these terms, triple bottom-line approaches to corporate social and environmental responsibility tend to presume either that subsystem and systemic sustainability are commensurable *a priori* of other considerations or that the market subsystem provides the basis for comparison and is always oriented as a contribution to the sustainability of the system overall. Understood in these terms, the triple bottom-line approaches presume that the environment is an externality that can be unproblematically internalized as an economic cost that requires no further debate in terms of the social consequences of such an 'accounting fix'. The burgeoning corporate social and environmental responsibility movement in this perspective is an example of cultural ideological holism. As the product of a synthesis between state engagement in a process of restructuring and decisive efforts to deregulate and privatize social relations, it contributes to the dismantling of dualism and, in particular, the social contract between capital and labour.[35]

This creates a problem for the advocates of social and environmental justice, insofar as major corporations that have been accused of pollution or labour exploitation, such as Unilever, General Electric, Shell, BP, Dow Chemical, Nike, Nestlé and pharmaceutical manufacturer Aventis, mining concerns Norsk Hydro and Rio Tinto and several major petroleum producers, are also well known as signatories to the global sustainability reporting codes, adopt the AA1000 standard or have developed 'in-house' approaches to stakeholder negotiations. Indeed, in 2006, General Electric, 'the world's largest maker of [nuclear] reactors', developed its 'Ecomagination marketing drive, which seeks to improve the environmental impact of all of GE's products, from locomotives to light bulbs, and plastics to water treatment plants'.[36] Similar moves have been made by Royal Dutch Shell and British Petroleum, firms that emphasize good stakeholder relations and represent themselves as green businesses engaged in stakeholder consultations to achieve sustainability, and by ExxonMobil, which for many years strongly resisted attempts by activists and its shareholders to 'foist' green initiatives upon it, but now incorporates formal procedures for dealing with stakeholders in order to promote social and environmental sustainability. In the wake of the global financial crisis, however, both Shell and BP have nonetheless reduced by relatively large sums their efforts to develop sustainable practices. As oil prices rose in 2008, for example, these firms redirected funds into more lucrative but also more polluting oil-extraction industries, such as the coal tar sands of northern Alberta, Canada.[37] Similarly, ExxonMobil has increased its efforts to develop hydraulic fracturing industries in Australia, Canada and the US, directing considerable resources towards media campaigns in those countries promoting 'jobs and the environment', while retailing corporation Wal-Mart has made a strong public commitment to greening operations while remaining resolutely opposed to implementing reform of its notoriously punitive workplace-relations regime.[38]

My point is not that these firms are engaged in subterfuge of one form or another. Rather, it is that when a socially and environmentally responsible firm goes 'beyond petroleum', commits to integrating environmental concerns into business decision-making while reducing spending on renewable energy sources, promotes the expansion of ecologically questionable nuclear power, demonstrably supports political corruption in the global South or resists living wage demands in the United States, asserting that these acts are problematic appears to set 'us against them'. Understood in these terms, the nexus between the new citizenship's rights to wellbeing and security from risk and duties to be self-responsible, on the one hand, and corporate social and environmental responsibility initiatives, on the other hand, does not nullify possibilities that existing social practices might be exposed as sources of injustice. What is reinforced, however, is a cultural presumption that the interests of all social actors – individual citizens and stakeholder communities, business, government and non-governmental organizations – coalesce around ambiguously defined holistic ideals. Indeed, some of its early academic proponents now regard corporate social responsibility as costly to implement and ineffective as a means for achieving sustainable development: a recent review of the practice suggests that 'most empirical research' has concluded that sustainability reporting efforts are ineffective and that 'there is no grand result' showing that it is necessarily beneficial to society. Moreover, these business researchers assess the overall social value of such initiatives to find that, primarily, 'firms are engaged in "rebranding" in order to get even more [profitable] mileage from their beyond compliance endeavours' and dealings with stakeholders.[39] That is, in almost endless rounds of stakeholder negotiations and forums aimed at 'win–win' outcomes, profit-oriented action too easily trumps social or environmental justice considerations as the key justification for action. Such research admonishes governments and non-governmental organizations for unduly 'heaping too much praise' on firms that practice corporate social and environmental responsibility because, by the 2000s, it had become 'difficult to find any decent-sized company that does *not* have some type of philanthropic program'.[40]

Similarly, many proponents of what is arguably the most influential of these approaches, the Global Reporting Initiative, have recently sought to rethink the non-contractual voluntarism central to its triple bottom-line ethos. Originating as a 'voluntary sustainability reporting initiative' used by businesses to report on the status of their activities to, and consult with, 'external stakeholders', the Global Reporting Initiative 'indicators' have become deeply influential in shaping and informing corporate-sponsored stakeholder forums and wider governance for sustainability efforts. Its advocates have noted that the Global Reporting Initiative and similar schemes have 'little room for applying quality control to the reports or the process used to produce them' because they are conceptualized around significant 'systemic problems':

> [A Global Reporting Initiative] sustainability report cannot be both comprehensive and simple; it cannot simultaneously meet all individual needs and expectations and be continuously evolving; the goal of efficiency and streamlining fundamentally competes with the goal of inclusiveness, egalitarianism and multi-stakeholder participation; and the goal of flexibility, evolution and participatory ownership of the Guidelines is inconsistent with the idea of mandatory participation. More fundamentally, the framing of Global Reporting Initiative as an efficiency gain and win–win solution to the shared problem of information management, while effective in mobilising wide participation in the formative stages, does not have the gravitas to create or sustain a social movement, as some had hoped.[41]

Where stakeholder participation takes place within the orbit of sustainability reporting frameworks such as the Global Reporting Initiative, these are unavoidably appended to profit-making strategies first and sustainable development second. This point has more recently become clear to the proponents of the Global Reporting Initiative. A 2009 consortium comprising the United Nations Environment Programme, the Global Reporting Initiative, global consultancy KPMG and the Unit for Corporate Governance in Africa retains the holistic metaphor yet describes what is required for the 'transition to sustainable markets and economies':

> The relationship between mandatory and voluntary approaches is framed differently today. Instead of presenting mandatory and voluntary sustainability reporting as exclusive options, they are in fact highly complementary. Assuming a complementary relationship between mandatory and voluntary approaches, the challenge for [national and supra-national] government then becomes to

> determine the appropriate minimum level of mandatory requirements. For the reporting entities [that is, market actors] the questions remains as to how much they would be prepared to do beyond their compliance with mandatory requirements.[42]

The normalizing of corporate social and environmental responsibility in the context of the destabilizing of the institutional arrangements of the welfare state obscures possibilities for debating the contribution of 'responsible' market actors to unsustainable development – that is, beyond the in-principle acceptance of expectations that citizens and communities always benefit from holistic approaches that rely upon localized non-contractual negotiations and non-territorial global conceptions of the production-consumption nexus. Meanwhile, even as the embrace by business of stakeholder-oriented corporate social and environmental responsibility has become a normal part of business activity since the early 1990s, it remains the case that carbon emissions and social inequalities have also increased over the same period.[43]

On the one hand, as stakeholders, citizens demand only consultation over how holistic ends may best be achieved. A kind of self-fulfilling prophecy has been established, wherein the holistic ends of the corporate social and environmental responsibility movement are taken as given, even while evidence grows that such approaches fail as contributions to sustainable development. A situation has emerged in which no one class or group in society benefits from unsustainable development, as did 'the bosses' from exploitation of 'the workers' in industrial society. When key political concerns are expressed from within a holistic cultural frame, what is at stake comes to be represented in ways that make it difficult to identify who benefits and who is responsible for wellness, be it cast in terms of present or future generations or of independence and autonomy in the context of risk. In the new era of capitalism that the corporate social and environmental responsibility movement closely associates itself with, all social actors are constituted as stakeholders, equally challenged by unsustainable development in *one world*. On the other hand, the empowerment of local communities and individuals in relation to the state and markets, as well as the rights and duties associated with it, does provide new opportunities to publicize 'the facts' about issues such as environmental pollution or climate change and to demand action not merely from the state but also from polluting businesses because they undermine rights to wellbeing and security from risk. That is, a new test of wellness has emerged from the dissolution of social citizenship, one that opens up opportunities for challenging injustice by referring to empirical evidence that corporate-sponsored stakeholder negotiations undermine opportunities for citizens to actively embrace social goods. In Chapter 6, I examine how some progressive social movements are challenging unsustainable development in these terms.

5.3 Democracy through the wallet?

Another aspect of social practice that has become increasingly normalized over the 1990s and into the 2000s is 'green consumerism', which is also closely linked with discourses of very weak ecological modernization and the new citizenship. Seen in relation to questions of citizenship, green consumerism is important insofar as it is often discussed as the grounding for a shift away from collective, solidary forms of social and political participation and towards the subpolitical forms of association that I discuss in Chapter 4. Insofar as the supply-side policymaking central to very weak ecological modernization disavows itself of the need for regulating demand, green consumerism rounds out a response to such policy by shaping the revealed preferences of consumers. It complements supply-side policymaking because, in green and, more generally, 'political consumerism, unfocused citizen–consumer concerns for sustainability are articulated, translated and directed to providers in production–consumption chains'. However, this is also said to give 'citizen–consumers authority and power in a non-trivial way'.[44] Green consumerism depends for its efficacy, in relation to the sustainability goal, upon the enlightened rational choices of consumer-citizens who enact 'democracy through the wallet' as they pursue subjective quality-of-life goals based in enlightened awareness of limited and maldistributed ecospace. As such, my point is that green consumerism, while having little, no or even negative impact upon ecological concerns, represents an institutional response that of itself dissolves the distinction between the private and public spheres. Governments, businesses, academics and some non-governmental organizations now regularly call on citizens to participate pro-actively in personal-use consumption as a way of 'engaging in sustainability'.[45] In this perspective, citizenship is premised upon types of social and political participation that support enlightened consumer sovereigns to 'act locally while thinking globally'. Contemporary citizens express their creativity and authentic individuality not only by choosing to contribute to innovative and responsible businesses as 'stakeholders' but also in choosing to consume green commodities and invest in green businesses. In this situation, citizens are called on to exercise consumer sovereignty through a privatized ethic of enlightened choice.

Indeed, it was noted by researchers and popular journalists in the early 2000s that increasing numbers of businesses across a range of industries, often supported by government and civil organisations, had taken to distributing goods and services by offering 'social messages'. These combine appeals to consumers' sovereign desires with appeals to concerns about social or environmental issues. Hence, calls for fundamental changes in lifestyle and consumption patterns that arose in the 1980s and 1990s prompted a wave of green consumerism. Meanwhile, the popular media regularly inform the public about their new responsibility to consume with green discrimination.[46] It is in these terms that public awareness of risk constitutes a target market for the purveyors of consumer items: citizens are called upon to independently participate in personal-use commodities markets as a way of engaging in the promotion of social goods. As such, consumers choosing one commodity over another are likened to voters choosing politicians in an election.[47] The policy assumption that self-responsible citizens will increasingly choose to consume with green discrimination also implies an assumption that trust, modelled on inter-personal bonds, is present. Indeed, in 'the globalizing of production–consumption networks, the role of *trust* (in distant providers, in invisible technologies, in complex and diverse information flows) gains special importance' in relation to the progressive claims made for green consumerism.[48]

In this sense, the sheer complexity of globalizing production-consumption chains, combined with the vicissitudes of markets and availability of information and counter-information, can also raise doubts about the capacity of individual citizens to effect substantive change. Put in darker terms, such are conditions in which 'neurotic citizens' are bound over to self-author narratives of resolution to pressing existential and societal problems. Viewed alongside rising emphases upon wellness, green consumerism responds to what Isin sees as citizen needs for 'soothing, appeasing, tranquilizing, and, above all, [self-]managing anxieties and insecurities', not the least of which are 'eco-anxieties'.[49] Such offers of opportunities to take part in challenging social and ecological injustice through pro-active participation in personal-use consumer goods and services markets coincide with the emphasis upon self-creativity and pro-activeness more generally. Where 'the creation of professional conditions for people [is] such that...their capabilities and economic needs are sufficiently assured to allow them to take initiatives and shoulder responsibilities',[50] enlightened citizens are effectively called upon to bear an ethico-moral burden for political problems within and beyond the state, and to assuage this burden by shopping.

In this sense, the ethico-political consequences of personal action are effectively sublimated into ethico-moral obligations that citizens exercise self-interest as green consumers. However, what is normalized is a kind of *ethics-lite*, whereby sovereign consumer choices – mediated through revealed preferences for 'green', 'environmentally friendly' or 'safe' commodities – express ethico-moral obligations that are devoid of obligations set by the political community. The new citizenship in this sense deepens the individuation of ethico-moral reflection while depoliticizing the production-consumption chains that, in practice, remain sources of the ethico-moral dilemmas that green consumption is set to resolve. Although many such campaigns go some way to alerting consumers about the politics of global consumption-production regimes, the point remains that personal acts of green consumption stand in for citizens' political obligations. In the place of calling upon citizens to engage in the political sphere, citizens are called upon to shoulder responsibilities alone. Genuinely green or merely greenwashing businesses 'perform' the ethico-moral obligation on behalf of citizens, who merely enact capacities for enlightened choices when shopping for the products of 'responsible' firms. Emphasizing the rewards of embracing personalized 'creative' consumption opportunities, the new citizenship privileges the aesthetic dimensions of social participation, as if political concerns were always elaborated into political acts directly as an experience of the subjective realm of authentic selfhood. Green consumerism provides self-responsible individuals with opportunities to improve their material life conditions and social status in ways that contribute to economic growth and may contribute to green outcomes: as a social good, sustainability is represented in holistic terms, '*as if* solutions to socially created problems were always synonymous with expedience in the private realm of sovereign choice'.[51] Stakeholder citizens are encouraged to be self-responsible in relation to pressing social issues, such as injustice, as if the personal choices made by them as consumers always and directly affected political practice. This tends to frame the considerations that provide motives for ethico-moral choices as gradations of quality and, as such, to represent efforts to assuage citizen awareness of socio-ecological problems in situations where political pressure points are diffuse. A key problem is that while citizens' ethico-moral commitments to just terms of participation within the ecosphere are deferred, these are, in the absence of legitimately authorized rules regarding trading relations, for example, subjected to pressure from structurally entrenched state imperatives to support capitalistic accumulation in ways that also privilege uninvolved or greenwashing competitors. Contemporary citizenship offers self-responsible individuals de-territorialized opportunities to address the impacts of their own actions upon others and within the ecosphere *as if* there were no legitimate authority charged with organizing the rules for life held in common.

For Kersty Hobson, policy for ecological modernization that aims to communicate the 'facts' of climate change to citizens with the objective of stimulating green-rational consumption choices also has unintended consequences. As Hobson shows, sustainable consumption strategies often complement 'eco-efficiency' drives, which promote to householders technological innovations such as energy-efficient light globes or water-saving showerheads. A key responsibility for citizens is therefore a ethico-moral one: to 'green' personal consumption choices while social institutions provide support for the continual enhancement of personal capacities for self-responsibility in relation to consumption choices. That is, the state and markets provide consumer information upon which it is assumed that enlightened citizens will act. However, as Hobson demonstrates, such interventions also tend to create 'discursive traps', whereby information supplied to individuals can have the effect of alerting them to, and prompting them to think through, the structural constraints upon achieving sustainable development.[52] In Chapter 6, I discuss the actions of the fair-trade movement that are addressing such issues by leveraging what Hobson sees as 'discursive traps'. This movement takes the emphasis of the new citizenship upon non-contractual and de-territorialized relations and the dissolution of the public-private division as a cue to call for fair trade regulation within and between states, in relation to both consumer goods markets and organizational procurement.

5.4 Capitalism and liveability

While some idea of liveability was popular at the end of the 19th century, when local leaders and real estate entrepreneurs used it to attract settlers to newly established towns in the western United States, the term was used to indicate something akin to 'quality-of-life' in the early 1970s, when cultural bodies, the media and urban planners adapted the concept to describe the agreeable lifestyle said to be offered by certain cities or regions. In 1972, the Greater Vancouver Regional District adopted the concept of 'liveability' as a core objective, and by 1977 the US National Endowment for the Arts had sponsored a series of 'liveability' programs through its Architecture and Environmental Arts Program, including the City Edges, Liveable Cities and Neighborhood Conservation projects.[53] Since the 1970s, the concept of 'liveability' has shaped public debate in these terms.[54] However, in the 1980s, a number of multinational consultancy firms began to develop liveability indices, especially in relation to large cities. The developers of such indices are quite specific about their aims and objectives. The well-known Mercer Index, for example, aims 'to help governments and major companies place employees on international assignments' while 'identifying the best infrastructure based on electricity supply, water availability, telephone and mail services, public transport provision, traffic congestion and the range of international flights from local airports'; similarly, the other well-known index, the Economist Intelligence Unit Global Liveability report, is designed for use 'by employers assigning hardship allowances as part of expatriate relocation packages'.[55]

Indeed, it is in part the popularisation of liveability as an ideal associated with quality-of-life and, in particular, wellbeing that has motivated the development and promotion by commercial organizations of various 'liveability indices' since the 1980s. The emergence and uptake by policymakers, politicians and planners of such indices is also closely linked with the shift to a postindustrial 'knowledge' or 'creative' society[56] that I discussed in Chapter 3. In this perspective, what the commercial liveability and sustainability indices offer is a technocratic determination of a socio-cultural issue with deep existential roots in modernist values of self-realization and holistic unity with the 'vital questions' of wellness. The presence of a tension between the technical rationality that the indices embody and the existential or socio-cultural rationality[57] supports the politically expedient use of the liveability indices. Such indices say little about issues that may sit beyond the scope of the data that they offer, such as economic distribution, discrimination, participation rates or the quality of representation or, indeed, societal impact within the ecosphere. This is a point that has been noted in a number of reports commissioned by cities often ranked highly in such indices.[58] Again, as in the case of corporate social responsibility, the indices create conditions for a self-fulfilling prophecy of sorts, one that sets its own measures of success while casting failure as failing to live up to the measures set by the prophecy itself. The key justice issue here is that the commercial indices are represented publicly *as if* the chosen indicators or benchmarks did not often conflict with broader conceptions of the common good. They do not necessarily indicate whether a highly ranked liveable or sustainable city or region is providing anything beyond what lay within the unilaterally determined list of indicators. Instead, the rankings are taken to be important by some political actors while demands for adequate monitoring of the impacts of a city's situation within the ecosphere is elided.

Put differently, the indices represent a form of technocratic over-reach that is complicated by the phenomena that they purport to measure: liveability or sustainability. The uptake of commercial liveability indices represents an institutional response to public demands for a more comprehensive understanding of the common good, one that calls into use various non-traditional metrics by and in cities. However, although drawing upon the techno-scientific research into quality-of-life in some cases, the commercial indices bowdlerize these attempts to measure quality-of-life. Early theories contrasted 'liveability' with 'comparison' theory. While liveability theory held that persons' judgements about quality-of-life referred to absolute standards or universalizable norms, comparison theory held that people make judgements about quality-of-life on the basis of comparison with some past experience, or with their own perceptions of the experiences of others.[59] In recent years, researchers have attempted to coalesce the two theories by proposing the view that persons' judgements about quality-of-life implicate both absolute standards and recent changes in quality-of-life.[60] In this view, approaching the problem of liveability or sustainability requires a combination of quantitative and, importantly, deliberative qualitative methods. However, this commercial use of the concept relies upon the idea that quality-of-life both is desired by the general public and provides a competitive advantage in conditions of globalization. In this usage, the concept emerges as a key discourse of legitimation for government and business. As such, quality-of-life is stretched to encompass concerns both with wellbeing or risk and with the surplus that is generated by economic growth, while being employed as a tool to promote a city, a region or in some cases a nation's attractiveness to global business or a transnational cosmopolitan 'creative class'. Hence, the liveability indices have a somewhat problematic relationship with quality-of-life, at least as it is discussed in the literature, and with sustainable development as a social practice. While liveability theory posits that people examine and assess their own experiences in relation to some absolute, such as poverty or health, or some universal norm, such as justice or fairness, the combining of such quality-of-life measures with comparative measures, such as cost of living, innovation or private school fees, in an index creates an epistemological tension. On the one hand, absolute or normative standards supply a concrete benchmark against which quality-of-life is measured and understood, even if comparative measures are included; on the other hand, the process of ranking relativizes judgement overall. That is, the commercial indices substitute an instrumental goal – climbing the index – for absolute standards or universalizable norms or for comparing quality-of-life as an aggregate representation of past personal experience or that of others.

This relativizing of judgement and the unresolved epistemological tension that is associated with it create a vacuum in terms of substantive judgements about liveability, sustainability or, as Alain Desrosières and others argue, any social problem that is subjected to the 'mania' for indices.[61] Into this vacuum fit discourses of unfettered access to consumer goods and services and the negative freedoms associated with mobility, creativity and authenticity. Where the commercial liveability and sustainability indices are used for political ends, a pre-determined holistic 'prophecy' stands in as the normative reference point for judgements relating to quality-of-life issues. When represented in the aggregate, such indices subsume what the academic literature regards as aspects of quality-of-life or, indeed, sustainable development, such as concerns with material distribution, social inclusion, democratic participation and, importantly, impact within the ecosphere, into instrumental goals such as climbing the index. It is in these terms, as a means for promoting a particular place with the aim of attracting international capital and capital-intensive 'creative' labour, that politicians and business actors popularize and promote the overall rankings provided by major liveability indices. Even though they include considerations such as the presence of museums, arts events and café culture, the commercial liveability indices offer a view of quality-of-life that is somewhat at odds with that promoted by the academic literature or, indeed, by the advocates of strong sustainable development. The commercializing of the concept of 'liveability' or 'sustainability' in some sense contravenes the epistemic and normative claims that have been central to the theorizing of quality-of-life and to the practical application of liveability theory through efforts such as the United Nations Development Programme's successive *Human Development Reports* or the Organization for Economic Co-operation and Development Better Life Initiative, which have their own, different shortcomings.[62] In Chapter 6, I discuss the movement to establish citizen-participatory social and environmental 'indicators of sustainable development' as one means by which an alternative to the political use of liveability and sustainability indices is being developed.

6
Justice after Dualism

A transformation of citizenship and, with it, the state and ideology has taken place in the West over the late 20th and into the 21st centuries. Central to this transformation has been a cultural ideological shift, one that has deeply affected possibilities for acting on criticisms of injustice. Concerns with 'distribution' that had been central to the prevailing model of social citizenship in the 20th century have been partly incorporated into relatively wider concerns. Contemporary actors involved in justifying a particular political ideology no longer do so against the backdrop of a dualist cultural ideology that sets socio-economic classes against each other in efforts to influence the rules for redistributing the wealth that subjugating nature had produced. The cultural ideological and the 'real' backdrop to political debate have shifted. Contemporary actors now justify themselves in relation to a holistic cultural ideology and postindustrial ecomodernizing state. This brings with it a new 'test' of the justice of political ideological claims, one that is cast against the backdrop of what the reflexive modernization theorists, such as Beck, regard as heightened awareness of 'risk' and subpolitical demands for autonomy. As social actors, parties bring to political debate a shared definition of the social good as 'wellness'. Whether a Latourian 'illusion' or not, modern dualism has been questioned by social actors and, to the extent that it was present, dissolved. In the West, the division between nature and culture and between the private and public spheres is blurred, the social contract between labour, capital and the state has largely been dissolved and social relations take place within and across de-territorialized yet ecologically finite local-global space. Moreover, in the context of a holistic cultural ideology, the legitimacy of Western state authority to organize the rules for life held in common has come to rest in demonstrating a capacity to support individual opportunities to take advantage of 'stakes' held in society. Such 'stakes' are defined in terms of wellness and turn upon discourses of independence in relation to quality-of-life and security from risk. These may but do not necessarily imply state support for progressive forms of justice. Moreover, a holistic cultural frame of reference for politics obscures possibilities for political debate that would support accusations that low quality-of-life and high levels of risk and dependency are consequences of the specific actions taken by particular groups or classes in society. That is, there appears to be no class or social fraction that can be directly held accountable for injustice in the postindustrial ecomodernizing state.

A post-dualist holistic cultural ideology has become established, and encompasses politics and political institutions, such as those of citizenship and the postindustrial ecomodernizing state that I focus upon in this book. This shift is in part the product of increased citizen awareness over the 1980s, and 1990s and into the 2000s and 2010s that something is drastically wrong with *how* society participates in nature. However, the transformation of cultural ideology from a dualist to a holistic frame, one that supports this formulation of the problem, has not taken place on terms of greens' choosing. Moving from dualism to holism, bridging the divide between the public and private spheres of action, supporting non-contractual and non-territorial social relations as well as ethico-moral awareness of the finiteness of the planet has not brought about a greener society. The currently unsustainable Western societies strongly promote private choices to consume with 'green' discrimination, but do not provide an adequate regulatory framework. Non-contractual civil law relations have become normalized, especially in relation to the occupational domain, while the prospects of obtaining a living wage from work have become slim for most citizens. The 'traditional' institution of the political community, that is, the state, is all too easily side-stepped in relation to its regulatory capacity to convert ethico-moral into political obligations in relation to ecospace. De-territorialized local-global relations have flourished in the era of networked governance, such that there seems no limit to the offshoring of jobs and environmental harms. Meanwhile, holistic corporate social and environmental sustainability initiatives and 'green consumerism' that have few, no or even negative discernible consequences for the distribution of ecospace flourish. That is, many of the hopes expressed by the normative theories have, in recent decades, been enlisted in the service of achievements that run contrary to the larger vision for justice that the theories of green citizenship propose.

Recently, it has been argued that what is needed to promote sustainable development is some form of green New Deal, supported by a greening of Keynesian economic policy.[1] Such a 'deal' would seek to replicate the welfare state compromise between markets, civil society and business that sustained both industrial development and the welfare state institutional arrangements of social citizenship in the 1950s and 1960s. However, as Dobson and David Hayes point out, such claims ignore the key condition for the redistributive compromise: the expansion of justice within such a frame was based in assumptions of an ever-expanding economy.[2] The aim of such a 'deal' would be to 'green' national and global economies and, in so doing, reduce both environmental harm, especially carbon emissions, and social inequalities within and between states. Yet, even if capitalistic economic growth is transformed into an engine for environmental reparations, such policy programmes neglect a key political condition of social citizenship: there is no social movement with the unitary vision or sufficient political power to push things in the direction of social justice, as the trades' union movement could, by withdrawing labour, within the industrial state. Moreover, in the era of globalization, transnational production-consumption chains, the financialization of markets and, of course, climate change itself create complex non-territorial interdependencies between states. These interdependencies serve to blur cultural, political, economic and ecological boundaries and, amongst other things, tend to bring states that uphold some degree of socio-environmental protection into competition with those that do not.

As such, even if adopting a stronger approach than is currently the case, the postindustrial ecomodernizing state would not directly address the tension between an economic system that is premised upon inequality between individuals and the necessary conditions for legitimation of the state by individuals, which, understood in terms of citizenship, encompass the provision of types of social and political participation, contents of rights and duties and institutional arrangements that promote equality, freedom and autonomy amongst citizens. This said, in the context of negotiations that continue to commit the international society of states to something resembling what Peter Newell and Matthew Paterson regard as a neoliberal 'climate capitalist utopia',[3] citizens are bound to force the issue of justice and, so, strong sustainable development onto the agenda, both within states and at the international scale. Forcing the issue of justice as sustainable development may or may not involve a green Keynesian New Deal. I now turn to reflect on the actions of a number of contemporary social movements by conceptualizing their efforts as working to reformulate the kinds of representation that can be made in the context of a holistic test of wellness. Given the ideological transformation and the 'greening' of citizenship and the state that has taken place over recent decades, I ask what political action to push states in this direction would look like if achieving justice requires explicit action by the state to uphold sustainable development.

My focus now turns to how a select group of movements – for environmental justice, financial transactions taxation, a guaranteed basic income, fair-trade regulation and social indicators of sustainable development – can be seen to evoke a holistic grammar in their demands for justice. I focus upon these movements and not others because they represent most clearly the consolidation of critique from within a holistic ideological frame. In many cases, such movements do not necessarily eschew dualism, the public-private split, contractualism or territorialism. However, neither do they 'hark back' to purely redistributive debates. This is not to say that actors, in particular social movements, are no longer interested in questioning the injustice of maldistribution of economic wealth as a consequence of class differences. Rather, it is to argue that such questions no longer appear publicly to represent the generalizable interests of all citizens. Moreover, such questions no longer gain traction in the political sphere. The movements that I discuss build upon the partial assimilation by the state and markets of the once-progressive critique of dualism to argue from within a holistic cultural ideological frame. Briefly examining them is helpful, insofar as it provides an opportunity to visualize what progressive claims for justice might look like *after* dualism. Other such movements might include the World Social and Thematic Social Forums, which are not in fact social movements but rather serve as focal points for networking and strategizing by social movements.[4] Similarly, the Occupy movement and Le mouvement des indignés have sought to galvanize a political will in the context of economic downturn in imaginative ways that are not expressed as calls for full-time employment but combine calls for redistributive justice with demands for meaningful economic activity.[5] The movement to incorporate the Capabilities Approach into environmental, ecological and sustainability economics, as a critique of one-dimensional neoclassical economic theory, calls upon an explicitly holistic metaphor of supporting social capabilities within ecological constraints.[6] The movement for a 'labour economy' that seeks to place 'labour at the centre of the economic process in harmony with nature' explicitly rejects the artificial transformation of autonomous labour power into the subjugated 'power of an automaton'.[7] Or, for example, the local currency movements work to shield localized producer-consumer relations from the unjust terms of trade that are dictated by economic globalization while eschewing economic or social separatism.[8]

These are movements that build upon the types, contents and arrangements of contemporary 'stakeholder' citizenship in order to assert their claims that injustice is present by evoking and demanding the expansion of the 'test' of wellness. Amidst a compromise on the social bond whereby the state is bound not only to provide external order and internal security, gather revenue, promote accumulation and maintain traditional legitimacy but also to manage the provision of 'stakes' in society, these movements are responding to, as well as working to modify, the kinds of claim that can be associated, via discourses of wellness and allied discourses of security from risk and human autonomy, with sustainable development. That is, these movements are seeking to redefine wellness by pushing the state to manage the tensions between collective welfare and individual interests in ways that promote a politically progressive definition of justice as sustainable development. The movements are exploiting the opening for critique that arises when it is recognized that wellness in all of its holistic dimensions – that is, of redistribution, recognition and representation or participation aimed at promoting subjective capabilities and societal capacities within the capacity of the ecosphere to provide for present and future generations – can provide the criteria by which status is judged and around which the postindustrial ecomodernizing state, acting domestically or internationally, is bound to manage the compromise on the social bond. The movements that I discuss here respond to the injustice of unsustainable development directly, using what I see as the prevailing grammar of justice in the ambiguously greening state. These movements are evoking the grammar of cultural holism while simultaneously eschewing a return to the grammar of dualism and, in doing so, establishing new progressive political discourses that translate ethico-moral ideas about the good life and how to treat others into political claims about justice.

6.1 Capable citizens and the environmental justice movement

Claims for environmental justice were initially noticed in the late 1980s, as 'a critical mass of community leaders, local activists, and businesspersons [began to work] with frontline staff of federal and state agencies and perhaps with others to address local issues that they care[d] about deeply'. The focus of this form of protest and negotiation was on spurring action by local and regional government to implement, or in some cases simply enforce, environmental regulation to protect communities from pollutants and other hazardous by-products of industry. Initially conceived as a response to technocratic 'top-down' government, this form of activism is closely tied to the political Third Way[9] and, indeed, what I see as stakeholder citizenship. Lamont C. Hempel describes such 'civic environmentalism' as a community-focused movement for justice:

> [that represents the] effort to focus sustainability strategies on the social, economic, and ecological well-being of communities. Participants in this movement define community sustainability in ways that highlight the relationships between local quality of life and local or regional levels of population, consumption, political participation, and commitment to intertemporal equity.[10]

In Hempel's analysis, the movement actively sought to situate claims for justice within a local-global frame of reference and to engage in local governance regimes. However, in the late 1990s and early 2000s, others such as DeWitt John argued that the initial movements' concentration on collaboration, compromise and efforts to define shared goals in relation to localized experiences of global problems had opened it to abuse by reactionary forces, which actually used the movement as 'cover' for efforts to do away with state-based environmental or, indeed, other social regulation.[11] In this view, even though civic environmentalism remains one avenue for politicizing unsustainable development, it can elide the presence of a disproportionate gap between the political power and cultural value-setting capabilities of the actors involved. It can favour market-oriented competitiveness between regions, cities and communities that reinforces rather than challenges unsustainable development.[12] In short, civic environmentalism can be understood in terms of the normalizing of stakeholder citizenship: demanding localized participation in the institutional arrangements of horizontal governance networks, and oriented to establishing rights to wellbeing and security from risk and obligations to be self-responsible for achieving these.

Indeed, as Julian Agyeman and Bob Evans and others argue, the civic environmentalism of the 1980s and 1990s demonstrated the critical potential of movements to address injustice in the ecomodernizing state.[13] However, these researchers identify the emergence of a 'stronger' critique out of dissatisfaction with civic environmentalism.

For Agyeman and Evans, the early Third Way-ist iterations of the movement merely express a 'narrow-focus civic environmentalism' that is primarily concerned with 'methodologically individualistic' conceptions of citizenship and with 'networking' to protect the environment, thus effectively divorcing environmental activism from wider questions of equity and governance. In the stronger view of what civic environmentalism had come to be concerned with by the late 1990s, being a citizen 'for' sustainability was being refashioned, such that it 'can only be understood as part of a reconstituted commitment to the processes of governance and justice'. In place of narrow-focus environmentalism and, specifically, of green citizenship itself, Agyeman and Evans see the environmental justice movement as opening up 'two different and complementary paths towards transformation'. This broader conception of the movement supplies a 'vocabulary for political opportunity, mobilization and action at the local level [and] a policy principle: that no public action [to address an environmental issue or risk] should disproportionately disadvantage any particular social group'. For Agyeman and Evans, what is important is that normative 'ecologism' and theories of green citizenship focus too narrowly on the environment and on ecological issues, therefore 'underplaying the broader social and political dimensions implicit in the concept of sustainability'.[14] In Agyeman's view, broad-focus civic environmentalism fosters a just sustainability paradigm, within which wider, structural issues of equality and fairness are made central to particular local debates over sustainability, liveability, quality-of-life, wellbeing and risk. As such, the environmental justice movement's localized assertions that injustice is present highlight the 'equity deficit' of citizenship as a lived experience, and the institutional arrangements and types of social and polticial participation that support this deficit.[15] The environmental justice movement asserts claims for particular social and environmental rights in the context of a socially unsustainable development trajectory. As such, and more importantly, broad-focus civic environmentalism treats unsustainable development as a holistic problem of human-to-human relations within the ecosphere. Rather than addressing injustice in dualist terms as a problem of redistributing the spoils of modernization as a social process aimed at subduing nature, the movement relates localized assertions that injustice is present to the global problems of reproducing social relations within the ecosphere. It demands that local communities have the capabilities and capacities to act on injustices that are a consequence of global risks.

Moreover, arising from within communities, the environmental justice movement situates assertions that injustice is present locally, while simultaneously bringing into view the macro-social structures that perpetuate injustice, and especially borderless global injustices associated with pollution or human and natural exploitation. As such, the movement gives expression to experiences that environmental and social justice issues are 'mutually constitutive'.[16] While stakeholder citizenship tends to level out differences by conflating winners with losers as together confronting risks, the environmental justice movement addresses injustice from the perspective of the 'level playing field' of the stakeholder society by operationalizing a holistic grammar. Hence, the actions of the environmental justice movement address the lack of a perpetrator problem without attempting a return to outmoded notions of class warfare, even though some movement activists and advocates may couch their actions in these terms. The movement addresses the perpetrator problem not primarily by challenging exploiters of humans or 'nature', but by challenging malfeasants for undermining the wellbeing of (poisoned worker or neighbourhood) citizens or for exposing them to risks that are directly related to 'the environment' in a particular place. Similarly, the environmental justice movement addresses the state as the legitimate authority charged with the task of organizing the rules for life held in common in a particular place. That is, claims for environmental rights of a specific kind in a particular place and covering a distinct group of persons are geared towards the state acting in this capacity, even when appealing to global norms or demanding inter-state action, such as at the United Nations Framework Convention on Climate Change meetings.[17] By promoting local and community self-responsibility to organize against structural sources of injustice, the movement asserts rights to wellbeing and to security from risk that push beyond conceiving of wellbeing as consumption-oriented 'choice' and self-responsibility for individually 'insuring' against risk. Insofar as a holistic cultural ideology has emerged to frame representations of society in terms of 'one world working together to achieve sustainable development',[18] the environmental justice movement redefines what it means to engage in politics by pointing to the minimum standards for justice that must be maintained by such a world. In situations characterized by types of social and political participation, contents of rights and duties and institutional arrangements that do not necessarily support progressive forms of justice, the environmental justice movement uses the holistic grammar that has become associated with the test of wellness to assert that injustice is present and that the state or inter-state institutions should act to uphold justice. In effect, the environmental justice movement exploits the partial and unsatisfactory greening of citizenship that was established in the 1990s around 'stakeholder' citizenship.

In working to define environmental justice, Schlosberg examines some of the implications of such actions. For him, the environmental justice movement shifts the focus of activism *per se*, from narrow concerns with the distribution of environmental goods to a broad, more comprehensive and politically legitimate concern with comprehensive rights to redistribution, recognition, representation or participation and, ultimately, 'capabilities'. While Schlosberg's argument that the environmental justice movement in fact 'extends [justice claims] to encompass the nonhuman world' may stretch things too far, the claim that the actions of the contemporary environmental justice movement expand the remit of environmental politics to encompass a politics of human-to-human relations within the ecosphere is demonstrable in the movements that Agyeman and Evans and others discuss. For Schlosberg, the environmental justice movement draws inspiration from an awareness that 'there is a relationship between everyday experience of disrespect, disempowerment, economic debilitation, and the decimation of individual and community capabilities and ... social movements such as civil rights, indigenous rights, gay and lesbian rights, feminism, postcolonialism and the more general movements for multicultural acceptance'. In this view, it is indeed the entry into political debate of the new social movements around environmental justice questions that has broadened and deepened prevailing ideas of justice. Hence, by 'encompassing the language ... [and] expanding the discourse of justice' the environmental justice movement speaks from within a general population that has become increasingly aware of and, in many cases, concerned with the impacts of environmental injustices or the 'risks' posed by pollution, climate change, nuclear power generation or resource over-use, for example.[19]

The environmental justice movement contributes to re-defining the role and function of the state and of justice within and beyond it because it appeals to and indeed emerges from a broader cultural ideological frame of reference. Rather than moving from an assertion about justice as the product of a particular dualist cultural ideological grammar, the environmental justice movement situates the experience of injustice as a consequence of the organization of human inter-relations as a whole within the ecosphere. It is in these terms that calls for equitable economic redistribution have become inseparable from questions of recognition, representation or participation and, ultimately, 'capabilities' that are intimately bound up with fair use of ecospace. The environmental justice movement draws upon the contents of stakeholder citizenship rights and duties to assert a distinctive claim for a socio-ecologically framed conception of justice, one that appeals directly to the test of wellness. The actions of the movement are both possible and effective because their claims are articulated within a wider cultural ideological transformation that demands an expanded political definition of injustice. The movement extends the ambit of the test of wellness and calls upon the state, and in some cases inter-state institutions, to implement policy that redefines what it means to hold a 'stake' in society. The environmental justice movement urges the state and inter-state institutions away from supporting policy for wellness that merely emphasizes consumer rights or voluntarism and towards support for 'capabilities', which are the products of adequate political representation and participation and equal recognition within ecological constraints that are defined by reference to claims for fair use of ecospace. That is, the progressive response to the stake-based form of citizenship that is central to the postindustrial ecomodernizing state is to be found in the environmental justice movement's demand for a capabilities-based citizenship that reformulates redistributive claims along holistic lines.

6.2 Challenging distribution: the movement for financial transactions taxation

The proposal for a tax on financial transactions, specifically to be levied by a state or states on foreign currency exchanges made within a particular jurisdiction, was designed by neoclassical economist James Tobin in the 1970s and modified in the 1990s by Paul Bernd Spahn.[20] Such a tax was initially to be levied at a low rate and to depend on the volume of trade passing through a particular financial market. The specific aim of such a tax was to cushion the extreme exchange rate fluctuations that became a normal feature of the international financial trade following the collapse of the fixed exchange rate mechanisms central to the Bretton Woods Agreements in the mid-1970s. Arguments that such a tax could actually reduce volatility in international financial markets remain open and rest in the presumption that reduced volatility is inevitably a social good. As such, the idea has been investigated or promoted by different British, Canadian, French and German governments as well as the G20 group and the European Central Bank. In 2011 a proposal for such a tax was ratified by all member-states of the European

Union, with the exception of the United Kingdom, as a response to rapidly deteriorating economic conditions within the Eurozone.[21]

However, claims that the resources generated by financial transactions taxation could be used to fund projects to increase social justice and/or environmental reparations only partly and tangentially rely on the presumption of stabilization. Beginning in the late 1990s, global justice movement activists and theorists argued that the revenues created by such a tax could and should be used to fund social and environmental justice projects, especially in the global South, primarily the Millennium Development Goals.[22] The central concern of global justice movement advocates for such taxation is, therefore, justice. Such advocacy is currently led by groups such as the Association for the Taxation of financial Transactions for the Aid of Citizens (ATTAC) and the Stamp Out Poverty[23] organizations, both of which have developed sophisticated programmes for implementing and administering financial transactions taxation in the form of a Robin Hood Tax. The global movement for a tax to be levied on financial services is interesting because of the way that calling for such a tax, whether at the global or national level, addresses the diffuseness of responsibility for injustice in the context of the postindustrial financialization of the economy. It is argued by its proponents that a financial services taxation regime conjures a relative scale – that is, a hierarchical relationship – of perpetrators of injustice. In the context of an emergent system of global climate capitalism that is 'creating markets, where money can be made for trading carbon allowances within limits set by governments… [a]gainst the backdrop of recalcitrant industries and reluctant consumers', the movement potentially identifies 'economic winners from decarbonisation'[24] and places them in a relationship with economic 'losers', who it is assumed do not currently benefit from financial transactions directly. That is, the movement for financial transactions taxation is important because it challenges the holistic framing of injustice in its own terms. Where many in the North and South benefit, albeit incrementally, from the involvement in financial markets of pension and sovereign wealth funds, for example, the movement seeks to attribute obligations to contribute tax receipts based on the volume of transactions.

Moreover, advocacy group ATTAC proposes extending the financial transactions taxation model to encompass a broader reform agenda that is specifically aimed at grounding justice in sustainable development and, in turn, something very close to the ideal of fair use of ecospace:

> Specifically, we fight for the regulation of financial markets, the closure of tax havens, the introduction of global taxes to finance global

> public goods, the cancellation of the debt of developing countries, fair trade, and the implementation of limits to free trade and capital flows ... [because] the preservation of our planet cannot be attained through technological progress and the commercialisation of natural resources, but ... requires a radical redefinition of economic development away from productivism and consumerism. This implies that the common goods of humanity (health care, education, water, climate, biodiversity, ...) must be given an international status that insures their protection, and that their preservation must be based on devoted financing through global taxes.[25]

If such recommendations were levied proportionally according to the degree of economic benefit gained (progressive in social terms) but also according to the degree of environmental harm created (progressive in environmental terms), this might have the effect of addressing unsustainable socio-ecological relations in a unique way. Profitable and harmful activities, such as currency trades, lending or share transactions that support oil exploration, would be penalized at a higher rate than profitable but beneficial activities, such as lending or share transactions for solar or wind power or public hospital construction, while those not profiting from either stand to gain a share of the funds generated by decarbonisation. In opposition to what Newell and Paterson regard as climate capitalist utopianism, efforts to develop and implement financial transactions taxation, especially in relation to carbon trading markets, make explicit the claim that without a tax, no profit from carbon should be available. The point that the movement for a green dividend from financial transactions taxation makes clear is that while it is agreed that 'internalizing externalities' is a non-negotiable objective, the net benefits of any activity associated with markets are always negotiable.

The movement thus shifts emphasis away from the critique of dualism from within a holistic perspective and towards a critique of holism after dualism. While potentially but not essentially global in scope and coverage, the movement urges action on what Paul G. Harris argues is a 'truly cosmopolitan global ethic' by creating political obligations that are placed upon 'better-off people everywhere'.[26] The objectives of the movement for financial transactions taxation have the aim of putting into action Elliott's argument that what is required for justice in the context of unsustainable development is an 'environmental harm convention':

> While not all of humankind contributes to environmental injustice, all of humankind does contribute in some way to environmental

> degradation, even if some do so disproportionately less than others. Some obligations, such as those to assist and compensate, may be unilateral but the commitment to a global environmental no harm principle – that is not to cause damage to the environment – must be universal.[27]

The movement for a green and possibly global regime of financial transactions taxation builds upon the types of social and political participation and contents of the rights of stakeholder citizenship and extends these in a number of ways. In particular, where stakeholder citizens are drawn into global financial networks, through voluntary or mandated participation in pensions and sovereign wealth funds, this movement aims to influence the ways that such finance is used. Where, as stakeholders, citizens' rights to wellbeing and security from risk are affected by global financial markets, the movement for financial transactions taxation aims to mitigate these impacts by re-defining economic development away from productivism and consumerism. Operationalizing the environmental harm convention in these terms implies a demand that 'winners' contribute, differentially and in direct proportion to 'losers', to supporting socio-economic fairness and socio-ecological mitigation and adaptation. That is, contra Dobson's argument from within the critique of dualism – that it is citizens in wealthy states who should accept the unilateral obligation to minimize their use of global ecospace[28] – linking financial transactions taxation with the environmental harm convention aims critique not at a dualist separation between rich and poor citizens or states but at the ideological holism which conflates all citizens as contributing in some way to environmental degradation. It recognizes that environmental degradation does not always entail injustice, but human exploitation does. This movement creates a political community that incorporates rights within society as a whole, yet applies responsibilities to citizens differentially. The practical outcome of enacting environmental harm conventions through financial transactions taxation would be to create holistic political obligations, insofar as winners and losers are identified and allotted a status within a possibly global political community. The beneficiaries of global financial markets, including those benefiting from generous pension plans or sovereign wealth funds in rich states or poor, would be obligated to provide restitution, while those who do not benefit – while in some sense still seen as 'contributing' to environmental harm – would not.

A second set of issues that the movement for financial transactions taxation addresses are those that Dryzek defines as 'constitutional excess' and 'excessive administration'. For him, non-contractual and non-territorial global relations seem to raise an unrealistic expectation that a liberal multilateral global constitution could be set in place without creating unwieldy administrative problems. Dryzek argues that, 'at the very moment administrative and constitutionalized problem-solving seem to be yielding to networked governance at national, regional and local levels, [some] still seek to risk replicating these questionable models at the global level'.[29] In light of Dryzek's two points, I contend that the movement for financial transactions taxation addresses these issues by promoting a global regime that can be formalized and administered from within a grouping or bloc of 'greening states' with the aim of both operationalizing environmental harm conventions and dragging recalcitrant laggard states towards sustainable development in a way that sets the burden upon the wealthy everywhere and lightens the burden carried by the poor anywhere. A precedent for such action is set by the European Union decision to impose a virtual 'border adjustment tax' on airlines arriving from nations without a carbon taxation levy in place from 2012 or recent, although not unproblematic, EU efforts to embrace a financial transactions tax. Although of course it creates its own plethora of political difficulties, I contend that it is not unreasonable to imagine financial transactions taxation avoiding the constitutional excessiveness and administrative unwieldiness that Dryzek associates with non-contractual, non-territorial global cosmopolitan aspirations. In short, such a facility would formalize 'advantaged citizen-[rich] state myopia and extend [cosmopolitan] arguments to their logical conclusion: *all* advantaged persons, regardless of whether they live in a [rich] state or a poor one, harm the world's poor'.[30]

6.3 Demanding recognition I: the movement for a guaranteed basic income

Advocated first in the 1960s and 1970s as an assertion of the right to a share in gross national wealth, the movement for a guaranteed basic income was somewhat devalued by the unmooring of the compromise on the social bond that had centred upon the test of wealth. More recently, the movement has been resuscitated. Practical proposals for how to implement a guaranteed basic income regime are currently discussed by movements such as the Basic Income Earth Network and supported by an inter-disciplinary academic journal.[31] The most widely advocated form of guaranteed basic income is based on the definition of Philippe van Parijs, who regards it as an unconditional monetary income regularly paid to individuals as members of a political community, irrespective of past or present wealth, household situation or employment.[32] Recommendations that it be funded in part by a financial transactions taxation regime or 'automated payment transaction' taxation, for example, are also popular in the literature and amongst activists, as are alternative approaches that advocate a combination of monetary and material 'stakes' in society such as housing.[33] Indeed, the recent claims differ from those made by the advocates of a guaranteed basic income in the 1970s, who saw it as a generic, non-contractual and non-reciprocal right of transfer between the capitalist and working classes. While in the 1970s such calls were largely the preserve of the welfare statist Left, contemporary calls for a guaranteed basic income are supported by those advocating political positions on both the Left, such as red-green coalitions, and the Right, such as libertarian individualists. Those on the contemporary Left tend to hold a civic republican perspective, and claim that a guaranteed basic income would provide all citizens with the means to avoid domination and potentially increase opportunities for citizenly political participation.[34] Meanwhile, supporters arguing from the Right regard it as a means for simplifying the taxation system, promoting responsibility for subjective fate and expanding negative freedoms.[35] Moreover, where citizens and communities are encouraged to be independent agents who assume self-responsibility for wellness – rather than receiving 'welfare' – the provision of a guaranteed basic income addresses the diffusion problem that I identify with responsibility for injustice in the ecomodernizing state without reverting to outmoded class antagonisms, which in any case lack political traction in the context of stakeholder citizenship.

Whereas the institutional arrangements of the welfare state promoted a view of social citizens as suffering from a structural condition, unemployment, the institutional arrangements of the postindustrial ecomodernizing state view stakeholder citizens as self-responsible actors who voluntarily choose to take up economic opportunities in order to build an income 'portfolio' from a range of sources. Similarly, social and political participation by social citizens was largely defined in terms of membership of mass-organizations such as trades' unions. In contrast, the types of social and political participation associated with stakeholder citizenship are defined in terms of voluntary contributions to localized community governance. Stakeholder citizenship in this sense creates a need to address the new 'luck egalitarianism', such that arguments in favour of a guaranteed basic income tend to emphasize its support for equality, freedom and, especially, 'autonomy objectives', such as exit rights from oppressive social relations, subjective free time and freedom from debilitating dependencies. Such proposals represent attempts to directly address the forms that injustice takes in the postindustrial ecomodernizing state, notably in relation to the deleterious impacts of the 'new worlds' of work and over-consumption. The guaranteed basic income, while possibly subjected to a means test to determine eligibility, in this sense breaks the link between social provision and full-time employment by focusing debates over justice upon income and 'portable rights' to remuneration for non-market activities, such as child-bearing or nurturance, rather than employment.[36]

Opponents of a guaranteed basic income, however, contend that a tension emerges in relation to calls for it and the demands of sustainable development, insofar as supply of the former would seem to depend upon continual economic growth. In response, its proponents contend that provision of a guaranteed basic income would support the cultivation of non-consumerist lifestyles while minimizing the share of commodities in aggregate consumption by citizens.[37] This is seen as feasible because, against the backdrop of unsustainable development, 'scarcity of land or jobs stems from ecological limits'; therefore it follows that 'the closer we [*sic*] get to these limits the more justified a [guaranteed basic income] would become, since people are not able to earn their necessary income through farming or wage labor'.[38] Attached to a green regime for administering financial transactions taxation as promoted by groups such as ATTAC, a guaranteed basic income can be understood as addressing the problems of diffuseness of responsibility for unsustainable development on the basis of what Elliott calls the environmental harm convention. Calls for a guaranteed basic income enunciate a demand that the 'requirement for personal responsibility [for ameliorating injustice] may be imposed only when the demands of equality are satisfied' and citizens are 'guaranteed a decent share of the social product'.[39] In the context of the postindustrial ecomodernizing state, such a claim points to a recognition that citizens bear 'common but differentiated' responsibility to contribute to steering this process by reducing their use of ecospace. Its advocates allege that the guaranteed basic income can potentially provide a grounding for citizen responsibilities to support sustainable development by promoting a net reduction in the use of ecospace, either by reducing consumption, decommoditizing social life or offering some form of what Barry calls 'sustainability service'.[40] Its advocates thus argue that a guaranteed basic income, one that is 'linked to (and supportive of) the expansion of community-based provision, volunteer work, cultural and sports activities, etc. [*sic*], could help offer more direct, resource-efficient and, thus, ecologically sustainable paths to well being'.[41]

That is, provision of a guaranteed basic income would contribute to decommoditizing social life and promote what Thomas Princen defines as 'sufficiency principles'. Indeed, as a policy proposal, the movement for a guaranteed basic income links itself to proposals for national or global financial transactions taxation and progressive income tax reform because it would contribute to socializing responsibility for sufficiency. In Princen's view, the sufficiency principles of 'restraint, respite, precaution, polluter pays, zero and reverse onus ... lend themselves to the sustainability goal by dealing directly with issues of criticality, risk export, and responsibility evasion' that too easily shape policy and practice when the focus is directed at principles of 'efficiency and cooperation'. That is, sufficiency principles address a situation in which the whole of society faces or creates too much risk *and* in which certain members or groups in a society are able to displace risk onto others or the ecosphere.[42] That is, goals of efficiency and cooperation that have emerged as central features of ecomodernization policy in the postindustrial state do not necessarily address the problems of injustice that are associated with unsustainable development. In this view, a guaranteed basic income regime is said to provide a means for reframing 'income' as part of a 'sufficiency-dividend' that by definition penalizes the relatively well-off while shielding those whose lives are conducted more or less in line with sufficiency principles. Its proponents argue that a guaranteed basic income regime would minimize the 'exportation' of risks to citizens and communities as a result of individual and family income volatility, while also promoting self-responsibility for wellbeing and security from risk. Indeed, this is the central idea behind the International Labour Organization's argument for a 'social protection floor' that provides '[c]oherence between social, employment, environmental and macroeconomic policies as part of a long-term sustainable development strategy'.[43] By calling upon those whose contribution to unsustainable development is greater than that of others to contribute part of their income to benefiting the social whole, this movement challenges injustice in the absence of a social contract between citizen-worker-consumers, markets and the state. By arguing for a guaranteed basic income or social protection floor that includes but is not limited to economic income, and as such represents a 'stake' in society, the movement seeks to establish a form of neo-contractual stability for citizens. The establishment of a guaranteed basic income linked with a regime of green financial transactions taxation addresses the levelling effect of contemporary citizenship, wherein employees, employers, offshore producers, managers, shareholders and, indeed, all are considered to exist as the holders of 'stakes' in society. Taken together, the movements for financial transactions taxation and a guaranteed basic income assert that injustice is a consequence of an unsustainable whole system of social reproduction, from which all potentially benefit and to which each should differentially contribute.

6.4 Demanding recognition II: the movement for fair-trade

The fair-trade movement began in the 1970s and 1980s with the aim of promoting justice in relation to international trade, especially in consumer goods:

> [The fair-trade movement promotes a] trading partnership, based on dialogue, transparency and respect, that seeks greater equity in international trade. It contributes to sustainable development by offering better trading conditions to, and securing the rights of, marginalized producers and workers – especially in the South. Fair Trade organizations (backed by consumers) are engaged actively in supporting producers, awareness raising and in campaigning for changes in the rules and practice of conventional international trade.[44]

In light of this definition, there are two dimensions to this movement. On the one hand are the individual or institutional consumers who 'back' the movement. While individual consumers do this by choosing to purchase goods or services that bear a mark of certification and, so, are recognizable as fairly traded, institutional actors that support fair-trade and ethical consumption movements do so voluntarily or are subjected to political pressure to make purchases or procurement arrangements that privilege fairly traded, locally manufactured or sourced goods or services. On the other hand is the movement itself, which works to promote the fair trade of goods and services within and between Northern and Southern nations and to coordinate production-supply with consumption-demand, usually but not exclusively between consumers in the North and producers in the South, and agitates for fair-trade procurement programmes and legislation or regulation for fair producer wages and conditions. I concentrate here upon the movement itself, and in particular on that faction or wing of the movement that actively pursues fair-trade regulatory regimes at the national and international scales. This said, since the 1990s the fair-trade movement has increasingly shown signs of a split with fair-trading networks. While the fair-trade movement has maintained agitation for increased regulation of trade, fair-trade networks have by and large sought accommodation with 'neoliberalism', in some cases arguing that fair-trade helps makes free trade work for the poor. In light of this split, the fair-trade movement for increased regulation, led by groups such as the European Fair Trade Association, the Fairtrade Labelling Organization and Oxfam International[45] and supported by major national and international trades' unions,[46] puts into practice what Iris Marion Young regards as the assumption of 'responsibility for injustice by virtue of [agents'] structural connectedness to it, even though [the movement and its members] are not to *blame* for it'.[47]

In this sense, it might be argued that the regulatory wing of the fair-trade movement enacts what Dobson regards as the similar 'thick cosmopolitan obligation'. For Dobson, the empirical developments associated with globalization 'make us more actively aware of our obligation[s]...turn[ing] the theoretical possibility of transnational obligation...into everyday reality'.[48] Others argue that participating in fair-trade, and ethical or green consumption more broadly, is a politicized act of connecting distant places and spaces.[49] Such acts are said to link individual ethico-moral choices with those affected by such acts, fostering the emergence of 'new political subjectivities and new forms of political representation'.[50] However, in contrast to Dobson and the other advocates of such a view, I contend that the action of the fair-trade movement to establish fair-trade regulation within and between states achieves thick cosmopolitanism not simply by raising awareness of obligations but by working to conjure a political community and political obligations out of non-contractual and non-territorial globalized social relations. That is, such undifferentiated claims for the politicizing aspects of fair-trade consumption seem a little overblown. What I contend is required to make thick cosmopolitanism politically meaningful is precisely the political community that stakeholder citizenship dissipates. What makes this order of claims problematic – whether couched in Dobson's terms as an exemplar of green citizenship or otherwise – is not least the subjective difficulty of distinguishing between goods and services produced or created as a consequence of green consumerism, on the one hand, and fair-trade, on the other hand. Moreover, in the face of deliberate attempts to obfuscate or greenwash such a distinction and in the absence of regulation – that is, political efforts to establish and organize the rules governing political obligation – the exercise of voluntary self-responsibility that such claims rely upon is effectively overburdened. In this view, the movement aspect of fair-trade, insofar as it attempts to entrench a politics of justice in relation to globalized trade relations or national 'neoliberal' regulation of workplace relations, actively transposes ethico-moral obligations into political ones through collective action in a political sphere that is experienced as something distinct and different from the private sphere of moral agency. The movement achieves this by explicitly recognizing and seeking to establish new institutional arrangements based upon the presence of hierarchical social relations, most often between the global (consumer) North and (producer) South but also within northern countries, especially the United States in relation to the low-wage and highly casualized service economy and calls for 'living wage ordinances'.[51]

It remains the case that the types of social and political participation that are enacted through the fair-trade movement create opportunities to act in 'ethical' ways for some and not all. That is, in the absence of all-encompassing regulation against unfair trade in goods and services, the presence in society of fair-trade goods and services creates an ethico-moral issue, insofar as they remain expensive for most consumers. For Clive Barnett and others, fair-trade 'is suggestive of new forms of practice through which unequal power relations are constituted, reproduced, and contested':

> This implies that there may be a basic contradiction between the means and ends of ethical consumption, in so far as the practical devices through which an ostensibly universalistic responsibility is made possible are also a means of socially and culturally differentiating certain classes of persons from others.[52]

However, when a distinction is drawn between the promotion of private ethico-moral choice by fair-trade networks and efforts by the fair-trade movement to promote political regulation, it becomes clear that while the former promote justice on the basis of an ethico-moral obligation, the latter promote justice on the basis of assuming political obligations. What the fair-trade movement does is make explicit the existence of hierarchy in the context of stakeholder citizenship, which flattens and relativizes the distinction between actions in the private and public spheres and obscures socio-economic tensions: the poor buy cheaper goods and services, while both green consumerist and fair-trade and ethical consumption goods and services tend to be more expensive. Understood in these terms, the fair-trade movement makes visible what

Young defines as 'structural connectedness' not by 'isolating' perpetrators according to guilt or liability but by creating an association between injustice and social structure, by acting on and judging social relations in terms of 'background conditions', by 'looking forwards' and challenging the persistence of injustice 'unless social processes change', by recognizing 'shared responsibility' for injustice as a generic quality of a society and, most importantly, by working to have that responsibility 'discharged only through collective action'.[53] That is, the fair-trade movement demands some form of regulatory requirement, one that by definition requires the political sphere that is currently grounded in the state to adequately recognize the wellness of others in producing communities.

In this sense, consuming individuals act out the existence of a hierarchy that is obscured at the level of private ethico-moral choice, while the movements that act 'behind the scenes' petition politically for that hierarchy to be recognized as a structural, whole-of-society instance of injustice. In effect, the movement motivates what Dobson and Barry[54] regard as 'the missing middle' of green citizenship as a practice. Dobson and Barry draw attention to the need for some kind of political community to act as both regulator and arbiter of the compromise on the social bond. While individuals may be 'acknowledging a responsibility to act to address wrongs for which one is not, in any causal sense, liable for or to blame',[55] citizens are not assuming responsibility for the creation of injustice unless also accepting the need for regulation of this responsibility. In this sense, it takes political action by a movement, such as the fair-trade movement, to render private subjective 'acknowledgement' effective as a contribution to debates over justice in the context of the test of wellness. Insofar as the fair-trade movement rejects green consumerism and seeks regulation and mandatory compliance schemes at the national scale and 'fair' over 'free' trade agreements at the international level, it acts to highlight differences between indiscriminate globalization and just internationalization, the wholesale integration of global markets and just terms for interdependence. The fair-trade and ethical consumption movements do this not so much by fostering what Dobson himself calls 'cosmopolitan nearness'.[56] Rather, the movements foster 'nearness' only by drawing attention to the obfuscation of the private-public split by green consumerism and 'neoliberal' networks that act on claims that fair-trade helps make free trade work for the poor. Indeed, the need for internationalization of national or supra-national blocs of interdependence in global trade is recognized by scholars working in the field of ecological economics as a key factor in establishing justice as sustainable development.[57] The fair-trade movement addresses the dissolution of the private-public split not by calling for its reinstatement but by recognizing that the ethico-moral obligation to assume responsibility for consumption choices also entails political obligations that build upon global 'structural connectedness' to recognize and formalize the status of all citizens as actors contributing to the reproduction of society within the ecosphere.

6.5 Rethinking representation and participation: social indicators of sustainable development

A key impetus for the localization of governance for sustainability is Section 28:3 of the United Nations' Agenda 21, developed and subsequently amended at the Earth Summits in Rio de Janeiro in 1992 and Johannesburg in 2002:

> Each Local Authority should enter into a dialogue with its citizens, local organisations and private enterprises and adopt 'a Local Agenda 21'. Through consultation and consensus-building, local authorities would learn from citizens and from local, civic, community, business and industrial organisations and acquire the information needed for formulating the best strategies.[58]

Another recommendation in Local Agenda 21, at Section 40:4, establishes the need for complementary 'sustainability indicators' or indicators of sustainable development. Indicators of sustainable development were originally conceived as a means for shifting the emphasis of national and international policy from gross domestic product and gross national product towards broader 'sustainable' measures of the common good.[59] Section 40:4 of Agenda 21 describes how indicators of sustainable development, it was hoped, would shift the focus of 'development' from a singular focus on linear growth to a multiple or compound focus on qualitative change:

> Commonly used indicators such as the gross national product (GNP) and measurements of individual resource or pollution flows do not provide adequate indications of sustainability... Indicators of sustainable development need to be developed to provide solid bases for decision-making at all levels and to contribute to a self-regulating sustainability of integrated environment and development systems.

In light of this situation, both governance for sustainability and efforts to develop alternative sustainability indicators have proliferated, with mixed results for each. Localized and consensus-driven governance networks have proliferated, to the extent that the shift from government to governance for sustainability is an entrenched feature of sustainability efforts. However, in contrast, and notwithstanding continued efforts to develop them, sustainability indicators are yet to achieve the same levels of influence.

This said, drawing direct inspiration from both Sections 20:3 and 40:4 of Agenda 21 are programmes to establish social indicators of sustainable development. While often influenced by global negotiations and inter-, quasi- or non-governmental organizations, such as the Millennium Development Goals or Organization for Economic Development Better Life Index,[60] such social indicators of sustainable development are neither top-down statist or even supra-state initiatives, nor are they fully techno-scientific efforts to monitor, assess and control social life or qualitative 'grassroots' actions. Social indicators of sustainable development are designed as policy advocacy and public tools that can 'capture and measure a particular aspect of sustainability policy in an easily communicated form, allowing monitoring and the subsequent "steering" of policy, whether by internal management or external political pressure'.[61] Although most often promoted by civil society and non-governmental organizations, most such efforts require some support and input from national, state/provincial and local/municipal government, not only (or necessarily) as a flow-on effect of the 'greening' of policy or in the form of funding, but also as sources of quantitative demographic and statistical data. Such support and information can also be provided by businesses or business groupings, especially insofar as social indicators projects are often the products of governance 'roundtable' discussions aimed at a particular aspect of sustainable development, such as urban planning issues. Many social indicators projects are tied up with universities and research institutes, which offer support for interpreting and managing the data. These also often provide the physical spaces for storing and updating the indicators, as well as for stakeholder meetings to determine the issues that should be measured and monitored. In particular, local collaborative and participatory social indicators of sustainable development have been widely researched, especially insofar as the initial enthusiasm surrounding them has waned and as state institutions have increasingly adopted them. The two most common approaches to understanding social indicators of sustainable development are governmentality and deliberative democratic theory.

Researchers such as Yvonne Rydin and Ted Rutland and Alex Aylett highlight some very important shortcomings of local collaborative social indicators of sustainable development projects. Such projects, it is claimed, lack influence over policy decision-making and fail to represent 'truly participatory planning'. Rydin, for example, argues that such projects can tend to concretize central-local relations around a responsibilizing of local government, even as local communities resist such responsibilization. As Rydin notes, this creates something of a methodological problem for governmentality approaches to social indicators projects. While recognizing that the approach is well suited to understanding the nature of 'central–local relations within contemporary governance', Rydin argues that governmentality offers less insight into 'the construction of subjects and objects to enable governance' in relation to defining or acting on sustainable development. As she suggests, the framing of analysis by governmentality theory, which emphasizes uncovering the capillary workings of biopolitics, makes it ill suited to dealing with the agency of actors who strategically use discourses and identify and exploit 'points of calculation' as tools for resisting and debating exploitation.[62] A further problem, I contend, is that governmentality theory also makes it difficult to distinguish between possibilities that some centralized, bureaucratically administered and techno-scientifically informed interventions – to set benchmarks against which the effects of policy can be reviewed, for example – may be normatively desirable for achieving sustainable development or, indeed, for facilitating community monitoring of policy. This can be the case when, for example, state/provincial or national government supports a social indicators project with the aim of monitoring and assessing the impact of local or municipal government policy.

Theories and practices of 'deliberative democracy' provide a counterpoint to one-dimensional theories of an all-pervasive governmentality. Indeed, a significant body of literature on local collaborative social indicators of sustainable development situates them as examples of deliberative democracy-in-action.[63] However, as normatively desirable as is the deliberative ideal of instituting a process through which open-minded actors are informed about a policy issue, consider its implications and complexities in non-coercive communicative terms and provide reasoned arguments for settling on a 'rational' course of action, some of its advocates recognize that in practice it can produce 'questionable developments for democracy'.[64] These include the use of deliberative forums to push specific interests and agendas, the emergence of a deliberative consulting industry and 'revolving door' between advocates and industry, and a tendency to sometimes deepen conflict between sectoral interests or to promote and inadequately deal with the tensions between technical and scientific issues or claims about 'sham' participation. Emphases on locality, collaboration and consensus in deliberative indicator selection projects can elide the presence of a disproportionate gap between the political power and cultural value-setting capabilities of the stakeholders involved in implementing action.[65] Indeed, Boltanski and Chiapello are also critical of the communicative model central to deliberative democracy *per se*, arguing at length that it mimics too closely the contours of the 'new spirit of capitalism'.[66]

However, social indicators projects are also forums in which the holders of interests beyond that of working *directly* to achieve sustainable development can agree upon a course of action and set benchmarking 'indicators' for implementing, monitoring and reviewing the products of that action in a particular place. This reveals an important tension between the emphases of governance on efficient cooperation and the capacity for localization, collaboration and consensus-building among 'stakeholders' to remain focused on the sustainability goal. Whereas corporate stakeholder engagement processes or commercial 'liveability' indices repackage political issues such as economic inequality or environmental impact as issues that can be addressed in the course of generating profitability for a business or apolitical techno-scientific matters, social indicators of sustainable development provide a basis for active participation in monitoring the impacts of corporations or governments. In this sense, social indicators can serve as 'catalysts' for establishing networks of trust between and across policy communities by constructing meaningful dialogues about the social dimensions of sustainable development that become embedded institutionally over time.[67] In this view, collaborating parties to such projects may not necessarily have as their own primary objective a direct orientation towards achieving sustainability. Rather, participants commit to collaborate on a working definition of sustainable development in relation to a particular place or project, while bound to their own sectoral interests or goals. The most obvious example is the business organization, which despite employing staff at all levels who may be enthusiastic about embracing sustainable development, is nonetheless structurally bound to maintain profitability. Other sectoral interests might include so-called NIMBY groups, as well as organizations such as trades' unions or religious groups, which hold a responsibility to members that is not necessarily congruent with sustainable development.

This deliberative dimension of social indicators projects means that the holders of sectoral interests are confronted with the views of others and agree to work together to define a problem and how to measure and assess it. Claims for a working definition of sustainable development that appeal to self-interest alone, for example, are readily exposed in the deliberative forum process as having few or low consequences for achieving the stated goal of such a forum. Localized participatory social indicators initiatives have the potential to contribute to 'secur[ing] high-quality deliberation, given that partisanship is an inherent part of political life'.[68] This is because the newer initiatives, such as the Australian Liveable and Just Toolkit or Canadian Vancouver Vital Signs,[69] when understood as responses to the shortcomings of guidelines such as Local Agenda 21, contribute to a situation in which partisans and non-partisans need to deliberatively arrive at 'factual' indicators, in order that all involved take the first steps towards ameliorating an unsustainable situation. These draw partisans and other actors bent on achieving instrumental goals into something bigger while drawing up boundaries within which deliberation can open-endedly focus upon the worth of social action for a distinct sustainable development issue or issues, such as household waste levels, access to green spaces or resources directed at public education. Research that has discussed localized collaborative, consensus-driven participatory social indicators projects suggests that these often make a positive contribution to citizen inclusion in debates over sustainable development, albeit in some cases noting the difficulties in motivating citizens to participate in the early stages of such projects.[70] Indeed, in his survey of several social indicators and similar projects in the United States, Clark A. Miller argues that they are 'modifying the civic epistemologies of democratic societies' and 'working out new arrangements for making public knowledge and connecting it to public decisions'.[71] Such projects aim to uncover and define the social impacts of unsustainable development and, thus, provide links between macro-institutional orientations towards providing wellbeing as a key imperative of the state and cultural demands for 'sustainability' that may or may not run congruent with the interests of all stakeholders. Other examples are the Sustainable Seattle Indicators in the US; the Community Accounts developed in Newfoundland, the Democratizing Data program in Ontario and the BC Community Indicators and Regional Vancouver Urban Observatory in British Columbia, Canada; the Community

Indicators Victoria project in Australia; and, the Citizens' Panels on Indicators of Community Involvement in the UK.[72]

By involving citizens in developing 'factual' evidence in order to define and monitor the social consequences of unsustainable development, participatory social indicators projects such as Sustainable Seattle and Community Indicators Victoria are limiting the range of claims that collaborators may table as justifiable contributions in consensus-based forums aimed at establishing 'indicators' of sustainable development. That is, some claims are measurably unworthy as contributions to achieving sustainable development. Thus, the open-ended governance that the social indicators movement pursues in fact establishes boundaries for collaborative governance by normalizing deliberation over categories of unsustainable policy and practice. Social indicators of sustainable development categorize action in relation to the test of wellness. Such projects contribute to formalizing and, so, reinforcing a situation in which 'diverse interests and perspectives [are] voiced on problematic situations'.[73] These 'partially decouple the deliberation and decision aspects of democracy, locating deliberation in engagement of discourses in the public sphere at a distance from any contest for sovereign authority'.[74] The social indicators movement 'decouples' deliberation from decision by grounding the former in formal debate over what is required of a society to achieve sustainable development, thus potentially binding representative officials to act in particular ways.

The movement to establish and uphold social indicators of sustainability contributes to the establishment of what Hajer characterizes as 'public policy function[ing] as *public domain*', wherein policy discourse 'is *constitutive*...of political community' because it drives formalization of measures of policy failure or success.[75] By pushing the boundaries of localized and participatory governance regimes towards defining what counts as a contribution to justice, social indicators projects encourage critical debate that redefines the contours of citizenly representation and participation in society. Local community efforts to develop social indicators enact self-responsibility for delimiting and categorizing what is acceptable and unacceptable as a contribution to achieving justice as a common project. Collaborative, indicators-oriented projects that bring individuals together to participate in designing social indicators encourage efforts to rethink subjective aspirations for 'unbounded' self-realization. They do this by creating a space in which individuals can develop an understanding of the empirical conditions that enframe wellness as a consequence of social reproductive processes taking place within the ecosphere. Within the context of a holistic cultural ideology, where conditions of low solidarity prevail and individual self-affirmation displaces opportunities to forge political allegiances based in class- or status-group norms, social indicators of sustainable development provide a mooring point against which qualitatively and quantitatively derived 'data-sets' can anchor collective representations of wellness.

7
Conclusion

In the preceding chapters, I have reflected on a situation in which normative aspirations for the greening of citizenship, which outlined a clear emancipatory agenda in the context of the industrial state, have seemingly come to assume a far more ambiguous meaning in the context of the postindustrial ecomodernizing state. As an alternative to normative theories of green citizenship, I have developed a critical, pragmatic and realist approach that has supported interpretation of the greening of citizenship as a partially successful project that can helpfully be understood in relation to broader new social and countercultural movements that spread across the West from the 1970s onwards. In this perspective, efforts to disrupt the binary oppositions of nature/culture dualism and the public-private split, as well as calls to move beyond social contractualism and state territorialism while embracing ethico-moral awareness of the finiteness of ecospace no longer necessarily support the political critique of injustice: the globally competitive, postindustrial ecostate and the transnational forms of 'finance' capitalism that it supports also embrace these norms. This transformation in the status of hitherto emancipatory aspirations has been conceptualized in terms of three dimensions of society: *ideology, the state* and *citizenship*. Reversing the order in which they have been addressed through the book, I first revisit here my focus on the impacts of cultural ideological transformation upon possibilities for defining justice and injustice in terms of progressive and reactionary political ideological positions. Second, I revisit some of the implications of my focus on the functional role of the state as the political community that legitimately manages the compromise on the social bond between individual autonomy and collective obligation. And, third, I return to citizenship, as that social institution which constitutively defines the status and practices of individuals in relation to the state and other social institutions.

In *ideological* terms, I have developed an alternative view of what green political theory most often defines as nature/culture dualism. Rather than focusing on dualism itself, I have used Dumont's argument that a historically distinct form of 'modern artificialism' emerged in the 17th and 18th centuries out of Judaeo-Christian and, in particular, Protestant culture, to challenge what he regards as traditional or cosmological holism. In this view, an important 'ideological' effect of modernization was the establishment of a tension between modern dualistic and modernist holistic cultural ideologies. While modern dualism implies a cultural response that represents the unsettling experience of modernization as an attempt to maintain and reinforce the coherence of the atomistic self as 'lord and master' of nature, modernist holism implies a cultural response that represents the experience of modernization in terms of a desire to 'be at home in the maelstrom' through self-exploration and self-realization, as one element of an all-embracing whole. This view of culture as ideological is helpful because it provides the basis for interpreting what is more commonly defined as ideological: a political position. That is, taking culture to be the practices, discourses and material objects that express commonalities and differences, continuities and discontinuities of meaning over time – Williams' shared structure of feeling – has provided an insight into the representational context that encompasses a society. Recognizing and defining a cultural ideological grammar as that which can be identified with the 'whole set of representations that facilitate social integration and identity preservation' in a given society thus implies recognizing the political ideological discourses that seek to unmask, distort or dissimulate certain representations as inconsistent or unjust. As such, I have taken the view that, encompassed within a discernible cultural ideological frame, political ideologies are associated with efforts within a society to justify particular discursive appeals to the state as that political community legitimately authorized to organize the rules for life held in common.

Seeing things in these terms is supported by my interpretation of the primary political tension of modernization – what Lukes calls the 'grand dichotomy' between reactionary and progressive positions – as one affecting political ideologies in cultural ideological context. I have interpreted this difference as a consequence of the radical transformation of what Biro defines as the basic tendency for any biological organism to seek to go on in a given context, which arose with the establishment of modernization as a historical process. I have argued that this 'instinct for self-preservation', once grounded by a holistic culture of embeddedness within a collective way of life that is attuned to maintaining cosmological order, gave way with modernization in the West to a dualist cultural ideology that sets individuals 'artificially' against an objectified universe of material forces. On the basis of this assertion, it has been possible to define Lukes' 'grand dichotomy' in terms of progressive discourses that express broad self-interest, and which imply critical thinking towards collective creation of a cooperative order in which all are treated as equals, and reactionary discourses that express narrow self-interest, and which imply a defensive effort to justify existing personal or group privileges. Importantly, in this view, *neither* modern dualistic *nor* modernist holistic cultural ideological grammars necessarily entail *either* progressive *or* reactionary political ideological positions. Both modern and modernist, that is, dualistic and holistic, cultural ideologies can support progressive or reactionary political ideologies. Modern cultural dualism encompassed contrasting political positions that sought to justify subjugating nature with the aim of protecting existing structures of inequality *or* extending equality, freedom and autonomy to all citizens. Similarly, in the early 21st century, holism, cut loose from its historical association with organicist justifications of hereditary aristocratic privilege, can encompass contrasting political positions that seek to justify an elitist 'nihilism of the happy few' *or* extending conditions that support self-realization as capabilities for all 'amidst the maelstrom'.

As such, I have characterized the mid-20th-century differences between conservative and social democrats, as with that operating between the advocates of civil and political reform and adherence to political tradition in the 18th and 19th centuries, as differences defined against an encompassing dualistic cultural ideology. This has made it possible to situate contemporary political differences between advocates of the Tea Party-style movements and the Environmental Justice or Global Justice/Alter-globalization movements, for example, as differences between reactionary and progressive political positions in the context of a holistic cultural ideology. More importantly, in this perspective holism does not necessarily entail support for what is sometimes labelled the green ideology of ecocentrism or ecologism. The latter are rather conceived of here as subsets of holism. Hence, what I have regarded as being essential to progressive demands for justice is not a commitment to a particular form of holism, such as ecocentrism, but rather a commitment to extending conditions for equality, freedom and autonomy to all citizens in the context of an enframing holistic cultural ideological grammar. That is, the theoretical perspective that

I have employed in this book is grounded by what Hayward defines as 'weak' anthropocentrism, which is for Dobson 'an unavoidable feature of the human condition'.[1] However, contra Dobson, who evaluates ecocentrism as the essential Utopian engine driving less radical anthropocentric environmental arguments,[2] the approach that I have adopted supports Goodin's contention that so long as the green movement acts as the political vanguard of ecocentrism, it faces an insurmountable obstacle because such a programmatic 'green theory of agency' cannot be derived logically or coherently from a 'green theory of value'.[3] It is in this sense that I have described a situation in which the central claims made in normative theories of green citizenship – to disrupt nature/culture dualism and the public-private split, move beyond social contractualism and territorialism and embrace ethico-moral awareness of finite ecospace – need to be understood as emblematic of new social and countercultural movement 'protest against the disruption of natural equilibriums' that had, in the 1970s for Dumont, 'for the first time in public opinion placed a check on modern artificialism' but which, today, appear neutral in relation to the unsustainable development trajectory of the postindustrial ecomodernizing state.[4] In short, counteracting nature/culture dualism has not directly fostered justice because holism is not a political but a cultural ideology that provides the shared representational grammar or encompassing context within which different political discourses might be evaluated as contributions to social life. The normalizing of holism has created new opportunities for progressive movements oriented towards achieving justice, just as it has fostered opportunities to define justice in reactionary terms of preserving existing structures of privilege.

In this sense, viewing *the state* in functional terms has meant drawing attention not only to the political but also to the cultural dimensions of what state theorists describe as historically cumulative 'imperatives'. In addition to the 'classical' imperatives of maintaining internal order and external security, revenue raising and capitalist accumulation, it was noted in the 1970s that an additional imperative of state had been established, whereby the very political 'legitimation' of the state by its citizens was seen to depend on the provision of social welfare transfers that redistributed the spoils of industrial production across society. In this view, demands for social welfare grew out of historically entrenched cultural expectations that society was rightfully engaged in a collective effort to completely dominate or subdue nature. Extending such an approach, I have argued that in the early 21st century a transformation in the legitimation imperative of state has taken place, away from regulating the redistribution of wealth and towards state provision of support for individual opportunities to take advantage of 'stakes' held in society. I have developed this contention by recourse to the theoretical conception of a permanent historical dialectic between *compromise* and *test*, in relation to which the state, acting as the authority charged by citizens with organizing the rules for life held in common, is bound to manage. Stopping short of recognizing what Boltanski, Thévenot and the others see as the 'multiple orders of justice' that coincide with societal modernization, I have instead revisited their work through that of Dumont, Gauchet and Lefort to interpret the functional role of the state as that of arbiter of the compromise on the social bond between individual autonomy and collective obligation. The state has been viewed in its capacity to act as the authority that ranks or defines the status of political assertions that injustice is present in the context of a distinct cultural ideology and 'real' social conditions. The state fulfils this role by referring to a historically contingent 'test'. This is not to argue that all in society want to achieve that which a test represents. Rather, a test provides the shared criteria for assessing the status of political decisions, such as government policy or citizenly rights and duties, in relation to the social whole. As such, I have argued that in the 17th, 18th and 19th centuries, the mercantile state fulfilled this arbitrating role by delivering 'policy' that recognized the high status of free and equal, autonomous individuals who, making the right ethico-moral choices, subjugated an objectified and alienated nature. And that by the 20th century, such a test had come to centre on the distribution of the wealth that was being produced by the industrial subjugation of nature. That is, the 20th-century industrial state was bound to act as arbiter of the compromise on the social bond and did so by balancing progressive demands for distributive justice based on equal membership of society with reactionary demands for distributive justice based on a principle of desert.

This perspective has been extended by recognizing that, against the backdrop of what I describe as the ideological shift from dualism to holism that accompanied the postindustrialization of society, the state has been called upon to arbitrate the compromise on the social bond by reference to a new test of *wellness*. A partial consequence of new social movement concerns to universalize expectations that life should be meaningful and grounded in self-reflection rather than in relation to a status group, as well as conducted in harmony with nature and others and more diffuse countercultural concerns with self-help, New Age and spiritual paths to self-realization, by the 1990s and 2000s the test of wellness came to provide the benchmark against which the postindustrial ecomodernizing state managed the compromise on the social bond. I have argued that accompanying the transformation of the legitimation imperative of state from a grounding in welfarist redistribution towards a grounding in the provision of opportunity-laden 'stakes' in society has been a turn away from policy defined in terms of 'simple' wealth and towards a broader set of objectives that encompass self-responsibility in relation to quality-of-life or security from risk – that is, wellness. This has meant recognizing a broadening of the terms on which the state must decide policy; to maintain legitimacy, the state must address quality-of-life, risk, trust and communitarian issues as well as social inclusion and participation and lifestyle concerns, amongst other things.

That is, the recrudescence of holism necessarily encompasses the possibility that social conditions are a consequence of society's participation in nature but only contingently encompasses progressive claims that the cause of social injustice is the arrangement of social reproduction around gross degradation of the ecosphere. In these terms, what is important for critical thinking about justice is not so much the holism or dualism of a particular representation but, rather, the relationship of a discourse or practice to reactionary or progressive political positions. If the state is to be regarded as culturally *and* politically legitimate in a holistic frame, it must support freedom, equality and autonomy in respect of wellness. It must provide a basis for ranking status in a 'real' situation – regardless of whether policy that settles the test of wellness is geared to reactionary or progressive political positions. In contrast to popular claims that the West has since the 1990s experienced 'the end of Left and Right' ideological divisions, I have evoked the motif of the holistic test of wellness to argue that this division has been deepened. Moreover, when cast in terms of the test of wellness, justice appears as a whole-of-society problem and as such has no identifiably responsible class of agents or *perpetrators*, and is *diffuse* in terms of duties to minimize it. On the one hand, reactionary claims that justice is served through the pursuit of narrow self-interest no longer need to appeal to secondary concerns with upholding tradition, particularism over universalism and essentialism over contingency. These claims merely refer to ever-greater negative freedoms that will further unburden deserving individuals and communities in their narrow, self-interested pursuit of security from risk through untrammelled consumerism. On the other hand, whereas progressivism had sought to universalize the possibility of freedom by equivalizing economic redistribution and providing 'social security', on the basis of the claim that all are free, equal and autonomous citizens of a political community that grounds itself in a state that is committed to subjugating nature, the social solidarity that underpinned such claims is no longer available. From within a progressive perspective, assertions that injustice is present have become more complex. Building on work by Schlosberg and Fraser, I have developed the view that progressive forms of justice in the early 21st century require that substantive calls for redistribution and demands for recognition of equal status and for adequate representation or effective participation be combined in ways that link it to consequential domestic and international policy supporting capabilities for human flourishing at the subjective level and societal capacities to contribute to the reproduction of global society in ways that remain within the constraints of the ecosphere. That is, progressive demands for justice in a holistic frame are demands that the state and international community respond by implementing multiscalar policies for sustainable development.

In constitutive terms, my focus has been upon delineating the contours of a form of contemporary 'stakeholder' *citizenship* that represents the partial and problematic achievement of what I identify as the five central claims in normative theories of green citizenship. I have developed a Marshallian understanding of the cumulative yet changing conditions, assertions and types of social and political participation, the contents of the rights and duties and the institutional arrangements that accompany historical transformation from the predominance of successive bourgeois, social and stakeholder citizenship forms. That is, I have outlined the contours of a form of citizenship that has arisen in part as a consequence of responses by the postindustrial state to citizenly assertions that the bureaucratic government, mass politics, industrial employment, environmental degradation and modern dualist culture that had sustained social citizenship and the industrial state are no longer politically legitimate or culturally acceptable. Through a series of examples, I have argued that the concerns of normative theories of green citizenship with eschewing dualism, unifying the private with the public sphere, with relations of trust across local-to-global communities and with ethico-moral awareness of the finiteness of ecospace can equally support a 'classless' one world-ism of excessive working hours, casualization and 'precariousness', narrow forms of participation in governance and instrumental networking, rampant yet ineffective green consumerism, greenwashing and astroturfing, an over-emphasis on the political power of private decisions and unquestioned acceptance of techno-commercial criteria for monitoring and assessing social policy. Meanwhile, at the global scale, a social non-contractual, de-territorialized global 'race to the bottom' in terms of social and environmental norms has emerged to support a political economy that provides an unfair share of ecospace to privileged citizens in both Northern and Southern states, whose pursuit of quality-of-life through markets and consumption is largely taken as given by the international community of states. That is, stakeholder citizenship is anchored by types of social and political participation that centre on self-responsible, independent individuals, acting locally through their 'stakeholder' communities. It modifies bourgeois and social citizenship rights by the addition of rights to quality-of-life and security from risk, while modifying bourgeois and social citizenship duties through the establishment of duties to be self-responsible for taking the opportunities that the state offers in the form of 'stakes' in society. Similarly, the institutional arrangements of the Rule of Law, representative government and the welfare system have been modified or, in the case of the latter, partially undermined by the institutional arrangements of stakeholder citizenship that are anchored in horizontal and networked governance regimes.

Of course, this is in no way an argument that normative theories of green citizenship are themselves responsible for supporting the injustices that I associate with unsustainable development within, or as an export of, the postindustrial ecomodernizing state. Rather, by steering away from normative theory and developing an alternative approach, I have sought to understand the greening of citizenship in terms that, as Boltanski and Chiapello contend, recognize that 'the price paid' by critique for being listened to is at least in part to see some of the values it has mobilized in opposing the state and markets over time 'being placed at the service' of these institutions in accordance with ongoing processes of ideological assimilation. On the basis of this analysis, I have argued that where the state is called upon to manage the compromise on the social bond by dealing with the problem of justice in holistic terms, as a problem of addressing the test of wellness, new opportunities arise for progressive movements. Acting in the context of a *holistic* cultural ideology, *postindustrial ecomodernizing* state and *stakeholder* citizenship, I have identified some progressive movements that can be understood as combining, to different degrees, elements of redistributive claims, claims for recognition of equal status and for adequate representation or participation. Moreover, I have found that these movements are extending the contents of justice while also expanding the 'frame' of justice to, not unproblematically, encompass the whole-of-society issues of promoting individual capabilities and increasing the capacities of both postindustrial ecomodernizing states and other societies to reproduce themselves within the constraints of a finite ecosphere.

As I have argued, the example of the environmental justice movement, by urging the state and international institutions away from supporting policy for wellness that merely emphasizes consumer rights or voluntarism and towards policy that enhances 'capabilities', demands adequate political representation and participation alongside equal recognition of political subjects that are defined by a distributive ideal that goes beyond 'income' to demand fair use of ecospace. Similarly, the movement for financial transactions taxation seeks to enact an environmental harm convention such that holistic political obligations are created which identify the different status of actors within the global political community. This movement challenges the postindustrial ecomodernizing state-sanctioned 'climate capitalism' that recognizes all citizens as contributors to environmental harm and, so, as holders of a stake in the ecosphere, but fails to adequately recognize the unequal status of individual contributions to 'environmental' harm. This movement seeks to extend what Harris defines as the cosmopolitan argument to its logical conclusion: *all* advantaged persons, regardless of whether they live in a rich state or a poor one, harm the world's poor. I have also argued that, by calling upon those whose contribution to unsustainable development is greater than that of others to contribute part of their income to benefiting the social whole, the movement for a guaranteed basic income challenges injustice in the absence of a social contract between citizens, markets and the state. By arguing for a guaranteed basic income or social protection floor that includes but is not limited to or conditional on economic income, and as such represents a 'stake' in society, the movement seeks to establish a form of neo-contractual support for citizens that recognizes their status as members of a society that by necessity utilizes ecological resources. Taken together, the movements for 'green' financial transactions taxation and a guaranteed basic income assert that injustice is a consequence of an unsustainable whole system of social reproduction, from which all potentially benefit and to which each should differentially contribute.

Insofar as the fair-trade movement rejects voluntarist green consumerism and seeks regulation and mandatory compliance schemes at the national scale and 'fair' over 'free' trade agreements at the international scale, I have found that it highlights differences between indiscriminate globalization and just internationalization, the wholesale integration of global markets and just terms for interdependence. The fair-trade movement address the dissolution of the private-public split not by calling for its reinstatement but by recognizing that the ethico-moral obligation to assume responsibility for consumption choices, to consume with 'social' or 'green' discrimination, also entails political obligations that exploit 'structural connectedness' to identify and recognize the status of all citizens as actors contributing to the reproduction of society within the ecosphere. In the final example, I argued that the movement to facilitate citizen participation in developing social indicators of sustainable development can help to categorize action through participation in relation to the test of wellness. Social indicators projects help to 'decouple' deliberation from decision-making by grounding the former in debate over what is required of a society to achieve sustainable development, thus potentially binding officials and businesspeople to act in particular, transparent ways.

Acting from within the West yet global in scope, these movements are establishing new progressive political discourses that translate culturally holistic ethical ideas about the good life and moral ideas about how to treat others into concrete principles that call upon the postindustrial ecomodernizing state, domestically and as a contributor to the international community of states, to support justice as sustainable development. These progressive movements are exploiting the opening for critique that arises when it is recognized that wellness, in all of its holistic dimensions – of redistribution, recognition and representation or participation in compatible terms that promote 'capabilities' and 'capacities' within a finite ecosphere – provides the criteria around which the postindustrial ecomodernizing state appears bound to manage the compromise on the social bond. These movements are effectively re-defining wellness by pushing the state to manage the tensions between individual autonomy and collective obligations in ways that promote progressive forms of justice. This is because states remain the key sources of legitimate political authority to organize the rules for life held in common amongst citizens who, as well as being political actors, share a cultural narrative that describes their belonging within the ecosphere and amongst each other. These contemporary progressive social movements are responding to the continuing reproduction and 'export' of injustice by postindustrial ecomodernizing Western states.

By way of final words, I contend that further research may find that contemporary progressive movements such as those examined here are enacting efforts to expand the contents of justice and to dispute the frame of justice in ways that demand that the postindustrial ecomodernizing state support certain concrete principles, domestically and internationally.[5,6] These I suggest might be defined in terms of *commonality, stability* and *formality*. *Commonality* may point to policy that enacts the understanding that the world's citizens share something, the ecosphere, that is both fragile and limited. In this sense, commonality resembles what Dobson regards as 'cosmopolitan nearness' or what, more recently, Barry describes as 'vulnerability and dependency',[7] with the important *caveat* that ethico-moral awareness of such relations is but one component of justice. To paraphrase Rosanvallon,[8] a good society is defined by the fact that it is instituted by its members, who share in something that endures in space and over time. Commonality would thus entail rights and duties that make the shared nature of social life within a finite ecosphere explicit. A concrete principle of *stability* may point to an understanding of the political community, be it the state or post-state institution, as the primary decision-making entity legitimately charged with organizing the rules for life held in common, over the long term and into an inherently uncertain future. Stability points to citizenly recognition of the necessity of establishing rules for life held in common that define not merely redistributive regulations but also those governing equality of status and transparent representation or effective participation and do not leave these to determination by 'free' markets, open-ended governance networks or voluntary compliance schemes, for example. And *formality* may point to an understanding of the political community – most often but not always, yet, currently, the state – as the primary 'agent of redress'[9] in cases of injustice, one that is guided by constitutional principles that are tempered by deliberative democratic participation, and which confront the informal arrangements of stakeholder citizenship. A concrete principle of formality may recognize that, whether in localized community networks or global trading relations, there is a need to settle ground rules, such as measures of 'liveability', fair-trade procurement regulations or provisions for a social protection floor, while providing the means by which citizens can regularly and formally appraise, review and ensure the implementation of such rules with a view to achieving genuinely strong sustainable development.

Notes

Preface

1. T.H. Marshall, *Class, Citizenship, and Social Development* (New York: Free Press, 1965); ——, *The Right to Welfare and Other Essays* (London: Heinemann, 1981).
2. B. van Steenbergen, 'Towards a Global Ecological Citizen', in *The Condition of Citizenship*, ed. B. van Steenbergen (London: Sage, 1994); A.P.J. Mol and G. Spaargaren, 'Environment, Modernity and the Risk-Society', *International Sociology* 8, no. 4 (1993); J. Barry, 'Sustainability, Political Judgement and Citizenship: Connecting Green Politics and Democracy', in *Democracy and Green Political Thought*, ed. B. Doherty and M. de Geus (London: Routledge, 1996); J. Burgess, C.M. Harrison, and P. Filius, 'Environmental Communication and the Cultural Politics of Citizenship', *Environment and Planning A* 30, no. 8 (1998); P. Christoff, 'Ecological Citizens and Ecologically Guided Democracy', in *Democracy and Green Political Thought: Sustainability, Rights and Citizenship*, ed. B. Doherty and M. de Geus (London: Routledge, 1996); A. Dobson, 'Ecological Citizenship: A Disruptive Influence?', in *Politics at the Edge: The PSA Yearbook 1999*, ed. C. Pierson and S. Tormey (New York: Political Studies Association/ Macmillan 1999); M. Hajer, 'Ecological Modernisation as Cultural Politics', in *Risk, Environment, Modernity*, ed. S. Lash and B. Szerszynski (London: Sage, 1996); B.S. Turner, 'Contemporary Problems in the Theory of Citizenship', in *Citizenship and Social Theory*, ed. B. S. Turner (London: Sage Publishers, 1993), 1–18; B.S. Turner, 'Outline of a Theory of Citizenship (originally published in *Sociology* 24(2): 189– 217)', in *Citizenship: Critical Concepts*, ed. B.S. Turner and P. Hamilton (London: Routledge, 1994 [1990]), 199– 226; U. Beck, *Ecological Politics in an Age of Risk* (Cambridge: Polity Press, 1995) ——, *Risk Society: Towards a New Modernity* (London: Sage Publishers, 1992).
3. Ibid., 4–5, 203.
4. van Steenbergen, 'Towards a Global Ecological Citizen', 147.
5. T. Hayward, *Constitutional Environmental Rights* (Oxford: Oxford University Press, 2005), 1.
6. See T. Hayward, *Political Theory and Ecological Values* (Cambridge: Polity, 1998); ——, 'Anthropocentrism: a Misunderstood Problem', *Environmental Values* 6, no. 1 (1997).
7. D. Schlosberg, *Defining Environmental Justice* (Oxford: Oxford University Press, 2007).
8. N. Fraser, *Scales of Justice: Reimagining Political Space in a Globalizing World* (New York: Columbia University Press, 2009), 15.
9. A. Sen, *The Idea of Justice* (Cambridge: Harvard University Press, 2009); M. Nussbaum, *Creating Capabilities: The Human Development Approach* (Cambridge: Harvard University Press/Belknap, 2011).
10. S. Vanderheiden, 'Two Conceptions of Sustainability', *Political Studies* 56, no. 2 (2008): 435.
11. R. Eckersley, *The Green State: Rethinking Democracy and Sovereignty* (Cambridge: MIT Press, 2004), 80.

12. L. Boltanski and E. Chiapello, *The New Spirit of Capitalism*, trans. G. Elliott (London: Verso Books, 2005 [1999]).
13. N. Fraser, 'Feminism, Capitalism and the Cunning of History', *New Left Review* II, no. 56 (2009): 97–8.
14. A. Honneth, 'Organized Self-Realization: Some Paradoxes of Individualization', *European Journal of Social Theory* 7, no. 4 (2004): 476.
15. D. Drache and M. Getler, eds., *The New Era of Global Competition: State Policy and Market Power* (Montreal: McGill-Queen's University Press, 1991); Eckersley, *The Green State: Rethinking Democracy and Sovereignty*.

1 Introduction: Citizenship, the State and Ideology in a Critical, Pragmatic and Realist Lens

1. A. Dobson, 'Citizenship', in *Political Theory and the Ecological Challenge*, ed. A. Dobson and R. Eckersley (Cambridge: Cambridge University Press, 2006), 224.
2. T. Hayward, 'Ecological Citizenship: Justice, Rights and the Virtue of Resourcefulness', *Environmental Politics* 15, no. 3 (2006): 441; A. Dobson, *Citizenship and the Environment* (Oxford: Oxford University Press, 2003), 89.
3. M. Wissenburg, *Green Liberalism: The Free and the Green Society* (London: UCL Press, 1998).
4. M. Wissenburg, 'Liberalism', in *Political Theory and the Ecological Challenge*, ed. A. Dobson and R. Eckersley (Cambridge: MIT Press, 2006), 31.
5. D.R. Bell, 'Liberal Environmental Citizenship', *Environmental Politics* 14, no. 2 (2005): 183.
6. M. Arias-Maldonado, 'The Democratisation of Sustainability: The Search for a Green Democratic Model', *Environmental Politics* 9, no. 4 (2000): 56.
7. J. Barry, 'Resistance Is Fertile: From Environmental to Sustainability Citizenship', in *Environmental Citizenship*, ed. A. Dobson and D. Bell (Cambridge: MIT Press, 2006), 23–5.
8. See also, P. Petit, *Civic Republicanism* (Oxford: Oxford University Press, 1997); J. Barry and K. Smith, 'Civic Republicanism and Green Politics', in *Building a Citizen Society*, ed. S. White and D. Leighton (London: Lawrence & Wishart, 2008); R.C. Paehlke, *Environmentalism and the Future of Progressive Politics* (New Haven: Yale University Press, 1989).
9. Dobson, *Citizenship and the Environment*, 50.
10. Ibid., 44.
11. Ibid., 46.
12. Ibid., 47, italics in original, 82.
13. T. Gabrielson, 'Green Citizenship: A Review and Critique', *Citizenship Studies* 12, no. 4 (2008): 430.
14. For an overview, see work by A. Biro, *Denaturalizing Ecological Politics: Alienation from Nature from Rousseau to the Frankfurt School and Beyond* (Toronto: University of Toronto Press, 2005); J.M. Meyer, *Political Nature: Environmentalism and the Interpretation of Western Thought* (Cambridge: MIT Press, 2001); V. Plumwood, 'Nature, Self and Gender: Feminism, Environmental Philosophy and the Critique of Rationalism', *Hypatia* 6, no. 1 (1991); ——, 'Inequality, Ecojustice and Ecological Rationality', *Ecotheology: Journal for the Study of Religion, Nature and Culture* 5, no. 6 (1999); R. Williams,

Problems in Materialism and Culture (London: Verso Books, 1989); G. Lloyd, *The Man of Reason: 'Male' and 'Female' in Western Philosophy*, ed. J. Ree, Ideas (London: Methuen, 1984).

15. Gabrielson, 'Green Citizenship: A Review and Critique', 435.
16. Throughout the book, I distinguish a social fraction – a disorganized but dissenting group within a larger one – from a political faction – a small, organized dissenting group within a larger one.
17. J.S. Dryzek and P. Dunleavy, *Theories of the Democratic State* (New York: Palgrave Macmillan, 2009), 250; R.E. Goodin, *Green Political Theory* (Cambridge: Polity, 1992).
18. T. Gabrielson and K. Parady, 'Corporeal Citizenship: Rethinking Green Citizenship through the Body', *Environmental Politics* 19, no. 3 (2010): 374.
19. Ibid., 375; Dobson, 'Ecological Citizenship: A Disruptive Influence?' 40–1.
20. Dobson, *Citizenship and the Environment*, 132.
21. J. Barry, *Rethinking Green Politics* (London: Sage, 1999), 99.
22. Dobson, *Citizenship and the Environment*, 133.
23. Ibid., 32–4, 99–106.
24. Hayward, 'Ecological Citizenship: Justice, Rights and the Virtue of Resourcefulness', 438.
25. A. Dobson, 'Citizens, Citizenship and Governance for Sustainability', in *Governing Sustainability*, ed. W.N. Adger and A. Jordan (Cambridge: Cambridge University Press, 2009), 136.
26. A. Dobson and D. Hayes, 'A Politics of Crisis: Low-Energy Cosmopolitanism', *OpenDemocracy* (2008), http://www.opendemocracy.net/article/a-politics-of-crisis-low-energy-cosmopolitanism; see also A. Dobson, 'Book Review: *The Politics of Climate Change*', *Environmental Politics* 19, no. 2 (2010): 311.
27. Barry, 'Resistance Is Fertile: From Environmental to Sustainability Citizenship', 28.
28. J. Barry, 'Review: *Denaturalizing Ecological Politics: Alienation from Nature from Rousseau to the Frankfurt School and Beyond* by Andrew Biro', *Environmental Politics* 17, no. 4 (2007): 689.
29. A. Dobson, 'Ecocentrism: A Response to Paul Kingsnorth', *OpenDemocracy* (2010), http://www.opendemocracy.net/andrew-dobson/ecocentrism-response-to-paul-kingsnorth.
30. Ibid., 2.
31. Meyer, *Political Nature: Environmentalism and the Interpretation of Western Thought*, 2.
32. Ibid., 5, 124, italics in original.
33. Biro, *Denaturalizing Ecological Politics: Alienation from Nature from Rousseau to the Frankfurt School and Beyond*, 159.
34. B. Latour, *We Have Never Been Modern*, trans. C. Porter (Cambridge: Harvard University Press, 1993 [1991]).
35. B. Latour, *Politics of Nature: How to Bring the Sciences into Democracy* (Cambridge: Harvard University Press, 2004), 37, 67, 87, 89.
36. I. Blühdorn, and I. Welsh, 'Eco-Politics Beyond the Paradigm of Sustainability: A Conceptual Framework and Research Agenda', *Environmental Politics* 16, no. 2 (2007): 193, italics in original.
37. For a full account of this argument, see I. Blühdorn, *Post-Ecologist Politics: Social Theory and the Abdication of the Ecologist Paradigm* (London: Routledge, 2000).

38. M. Shellenberger and T. Nordhaus, *The Death of Environmentalism: Global Warming Politics in a Postenvironmental World* (Washington: Environmental Grantmakers Association/The Breakthrough Institute, 2004), 5, 6, italics in original.
39. M. Shellenberger and T. Nordhaus, 'Evolve: The Case for Modernization as the Road to Salvation', in *Love Your Monsters: Postenvironmentalism and the Anthropocene*, ed. M. Shellenberger and T. Nordhaus (Washington: Breakthrough Institute, 2011), 14, 22–3.
40. E. Swyngedouw, 'Impossible "Sustainability" and the Postpolitical Condition', in *The Sustainable Development Paradox: Urban Political Economy in the United States and Europe*, ed. R. Kreuger and D. Gibbs (New York: The Guilford Press, 2007), 25–6.
41. Ibid., 33, see also 38.
42. Blühdorn and Welsh, 'Eco-Politics Beyond the Paradigm of Sustainability: A Conceptual Framework and Research Agenda', 199.
43. B.S. Turner, 'Contemporary Problems in the Theory of Citizenship', in *Citizenship and Social Theory*, ed. B.S. Turner (London: Sage, 1993), 3.
44. J. Valdivielso, 'Social Citizenship and the Environment', *Environmental Politics* 14, no. 2 (2005): 241–2.
45. E.F. Isin and P.K. Wood, *Citizenship and Identity* (London: Sage, 1999), 4, italics in original.
46. B. van Steenbergen, 'Towards a Global Ecological Citizen', in *The Condition of Citizenship*, ed. B. van Steenbergen (London: Sage, 1994); B.S. Turner, 'Outline of a Theory of Citizenship (originally published in *Sociology* 24 (2): 189–217)', in *Citizenship: Critical Concepts*, ed. B.S. Turner and P. Hamilton (London: Routledge, 1994 [1990]); ——, 'Contemporary Problems in the Theory of Citizenship'; ——, 'The Erosion of Citizenship', *British Journal of Sociology* 52, no. 2 (2001): 189–209; ——, 'Marshall, Social Rights and English Identity', *Citizenship Studies* 13, no. 1 (2009); Isin and Wood, *Citizenship and Identity*; Valdivielso, 'Social Citizenship and the Environment'; A. Latta, 'Locating Democratic Politics in Ecological Citizenship', *Environmental Politics* 16, no. 3 (2007); S. Susen, 'The Transformation of Citizenship in Complex Societies', *Classical Sociology* 10, no. 3 (2010).
47. Valdivielso, 'Social Citizenship and the Environment', 244; see also Latta, 'Locating Democratic Politics in Ecological Citizenship' 389.
48. Turner, 'Outline of a Theory of Citizenship (originally published in *Sociology* 24 (2): 189–217)', 193.
49. Susen, 'The Transformation of Citizenship in Complex Societies', 260.
50. Ibid., 263; Turner, 'Contemporary Problems in the Theory of Citizenship', 3; ——, 'Outline of a Theory of Citizenship (originally published in *Sociology* 24 (2): 189–217)', 202.
51. J.S. Dryzek, *Deliberative Democracy and Beyond: Liberals, Critics, Contestations* (Oxford: Oxford University Press, 2000), 83.
52. See T. Skocpol, *States and Social Revolutions* (Cambridge: Cambridge University Press, 1979); Dryzek, *Deliberative Democracy and Beyond: Liberals, Critics, Contestations*; C. Offe, *Contradictions of the Welfare State* (Cambridge: MIT Press, 1984).
53. Eckersley, *The Green State: Rethinking Democracy and Sovereignty*, 63.

54. R. Inglehart, *The Silent Revolution: Changing Values and Political Styles Among Western Publics* (Princeton: Princeton University Press, 1977); ——, *Culture Shift in Advanced Industrial Societies* (Princeton: Princeton University Press, 1990); ——, *Modernization and Postmodernization* (Princeton: Princeton University Press, 1997).
55. D. Schlosberg, *Defining Environmental Justice* (Oxford: Oxford University Press, 2007); ——, *Environmental Justice and the New Pluralism: the Challenge of Difference for Environmentalism* (Oxford: Oxford University Press, 1999).
56. Susen, 'The Transformation of Citizenship in Complex Societies', 271, italics removed.
57. Ibid., 273.
58. Latta, 'Locating Democratic Politics in Ecological Citizenship', 389.
59. Ibid., 391.
60. Schlosberg, *Defining Environmental Justice*, 8.
61. L. Dumont, *From Mandeville to Marx: The Genesis and Triumph of Economic Ideology* (Chicago: University of Chicago Press, 1977), 18–19. Cited in E. Chiapello, 'Reconciling the Two Principal Meanings of the Notion of Ideology: The Example of the Concept of the "New Spirit of Capitalism"', *European Journal of Social Theory* 6, no. 2 (2003).
62. R. Williams, *The Sociology of Culture* (Chicago: University of Chicago Press, 1995).
63. L. Boltanski and L. Thévenot, *On Justification: Economies of Worth*, ed. P. DiMaggio et al., trans. C. Porter, *Princeton Studies in Cultural Sociology* (Princeton: Princeton University Press, 2006 [1991]).
64. Chiapello, 'Reconciling the Two Principal Meanings of the Notion of Ideology: The Example of the Concept of the "New Spirit of Capitalism"', 155–6.
65. L. Dumont and C. Delacampagne, 'Louis Dumont and the Indian Mirror', *RAIN* 43 (1981): 6.
66. L. Dumont, *Essays on Individualism: Modern Ideology in Anthropological Perspective*, English ed. (Chicago: University of Chicago Press, 1986 [1983]), 217. Chapter based on seminar paper, 'The Anthropological Community and Ideology', presented at l'Ecole des Hautes Etudes en Sciences Sociales and the Department of Anthropology University of Chicago, and originally published in *L'Homme*, 18, no. 3/4 (1978): 83–110. Dumont argues that anthropology has not recognized the importance of such 'protest' to social formation.
67. Isin and Wood, *Citizenship and Identity*, 4.
68. Eckersley, *The Green State: Rethinking Democracy and Sovereignty*, 63.
69. L. Thévenot, 'The Plurality of Cognitive Formats and Engagements: Moving between the Familiar and the Public', *European Journal of Social Theory* 10, no. 3 (2007): 411.
70. Boltanski and Thévenot, *On Justification: Economies of Worth*; ——, 'The Sociology of Critical Capacity', *European Journal of Social Theory* 2, no. 3 (1999): 360.
71. The key models of justice developed by Boltanski and Thévenot are the *market* (valuing self-interested performance); *industrial* efficiency (valuing technical competence and long-term planning); *civic* engagement (valuing technique); *domesticity* (valuing local and personal embodied ties); *inspiration* (expressed

in creativity and valuing emotion or religious 'grace'); and, *renown* (valuing pubic opinion, charisma and fame). Further work by Boltanski and Chiapello identifies with contemporary postindustrial conditions a *projective/connexionist* model (valuing flexibility, spontaneity and networking). Meanwhile, alternate contemporaneous research by Thévenot with Claudette Lafaye and Michael Moody identifies a *green* model of justice (valuing wilderness and sensitivity to environmental limits). See Boltanski and Thévenot, *On Justification: Economies of Worth*; L. Boltanski and E. Chiapello, *The New Spirit of Capitalism*, trans. G. Elliott (London: Verso Books, 2005 [1999]); L. Thévenot, M. Moody, and C. Lafaye, 'Forms of Valuing Nature: Arguments and Modes of Justification in French and American Environmental Disputes', in *Rethinking Comparative Cultural Sociology*, ed. L. Thévenot and M. Lamont (Cambridge: Cambridge University Press, 2001).

72. Thévenot, Moody, and Lafaye, 'Forms of Valuing Nature: Arguments and Modes of Justification in French and American Environmental Disputes', 236.
73. P. Lascoumes, 'Les Instruments D'action Publique, Traceurs De Changement: L'exemple Des Transformations De La Politique Française De Lutte Contre La Pollution Atmospherique (1961–2006)', *Politique et Societies* 26, no. 2–3 (2007); ——, *L'eco-Pouvoir, Environnements Et Politiques* (Paris: La Découverte, 1994).
74. Ibid., 374.
75. Ibid., 17–18; Boltanski and Chiapello, *New Spirit of Capitalism*, 33.
76. M. Steger, *The Rise of the Global Imaginary* (Oxford: Oxford University Press, 2008).
77. A. Honneth, 'Organized Self-Realization: Some Paradoxes of Individualization', *European Journal of Social Theory* 7, no. 4 (2004): 476; R. Sennett, *The Culture of the New Capitalism* (London: Yale University Press, 2005); Z. Bauman, *The Individualized Society* (Cambridge: Polity Press, 2001).
78. Z. Trachtenberg, 'Complex Green Citizenship and the Necessity of Judgement', *Environmental Politics* 19, no. 3 (2010): 350.
79. R. Geuss, *Outside Ethics* (Princeton: Princeton University Press, 2005), 226.
80. R. Geuss, *Philosophy and Real Politics* (Princeton: Princeton University Press, 2008), 9–16.

2 Modern Artificialism: An Alternative Perspective on Nature/Culture Dualism

1. C. Mukerji, *From Graven Images: Patterns of Modern Materialism* (New York: Columbia University Press, 1983), 8, 11.
2. C. Mukerji, *From Graven Images: Patterns of Modern Materialism* (New York: Columbia University Press, 1983), 15, 19.
3. Ibid., 10–11, 15, 21.
4. M. Weber, *From Max Weber: Essays in Sociology*, trans. H.H. Gerth and C.W. Mills (New York: Oxford University Press, 1958 [1946]), 339.
5. P. James, *Nation Formation: Towards a Theory of Abstract Community* (London: Sage Publications, 1996); ——, *Globalism, Nationalism, Tribalism: Bringing Theory Back In* (London: Sage Publications, 2006).

6. L. Dumont, *Essays on Individualism: Modern Ideology in Anthropological Perspective*, English ed. (Chicago: University of Chicago Press, 1986 [1983]), 10; ——, Dumont, *From Mandeville to Marx: The Genesis and Triumph of Economic Ideology* (Chicago: University of Chicago Press, 1977).
7. Dumont, *Essays on Individualism: Modern Ideology in Anthropological Perspective*, 55–6.
8. See Weber, *From Max Weber: Essays in Sociology*, 339.
9. Dumont, *Essays on Individualism: Modern Ideology in Anthropological Perspective*, 56.
10. L. Dumont and C. Delacampagne, 'Louis Dumont and the Indian Mirror', *RAIN* 43 (1981): 6.
11. M. Gauchet, *The Disenchantment of the World: A Political History of Religion*, ed. T. Pavel and M. Lilla, trans. O. Burge, New French Thought (Princeton: Princeton University Press, 1997 [1985]), 68.
12. Dumont, *Essays on Individualism: Modern Ideology in Anthropological Perspective*, 56.
13. Ibid., 77, 237.
14. Gauchet, *The Disenchantment of the World: A Political History of Religion*, 188.
15. Ibid., 94.
16. A. Biro, *Denaturalizing Ecological Politics: Alienation from Nature from Rousseau to the Frankfurt School and Beyond* (Toronto: University of Toronto Press, 2005), 159.
17. T. Adorno and M. Horkheimer, *Dialectic of Enlightenment* (London: Allen Lane Publishers, 1973 [1947]).
18. Dumont, *Essays on Individualism: Modern Ideology in Anthropological Perspective*, 56, 217.
19. A. Supiot, *Homo Juridicus: On the Anthropological Function of the Law*, trans. S. Brown (London: Verso, 2007 [2005]), 14.
20. Gauchet, *The Disenchantment of the World: A Political History of Religion*, 94; M. Gauchet, B. Renouvin, and S. Rothnie, 'Democracy and the Human Sciences: An Interview with Marcel Gauchet', *Thesis Eleven* 26 (1990 [1988]): 147.
21. T.H. Marshall, 'Citizenship and Social Class', in *The Citizenship Debates: A Reader*, ed. G. Shafir (Minneapolis: University of Minnesota Press, 1998 [1950]), 94.
22. See B.S. Turner, 'Outline of a Theory of Citizenship (originally published in *Sociology* 24 (2): 189–217)', in *Citizenship: Critical Concepts*, ed. B.S. Turner and P. Hamilton (London: Routledge, 1994 [1990]), 194.
23. Dumont, *Mandeville to Marx*, 54, 80, parentheses, italics in original.
24. See H. Arendt, *The Human Condition*, 2nd ed. (Chicago: University of Chicago Press, 1958), 61–2, 253.
25. J. Habermas, *The Structural Transformation of the Public Sphere: An Inquiry Into a Category of Bourgeois Society*, trans. T. Burger and F. Lawrence (Boston Polity Press/Massachusetts Institute of Technology, 1989 [1962]), 111–12.
26. N. Abercrombie, S. Hill, and B.S. Turner, *Sovereign Individuals of Capitalism* (London: Allen & Unwin, 1986).
27. Habermas, *The Structural Transformation of the Public Sphere: An Inquiry into a Category of Bourgeois Society*, 25–6, 102.

28. M. Gauchet, 'Democratic Pacification and Civic Desertion', *Thesis Eleven* 29 (1991): 10.
29. M. Gauchet, *The Disenchantment of the World: A Political History of Religion*; C. Lefort, *The Political Forms of Modern Society: Bureaucracy, Democracy, Totalitarianism* (Cambridge: MIT Press, 1986); C. Castoriadis, The Imaginary Institution of Society, trans. K. Blamey (Cambridge: Polity Press, 1987 [1975]).
30. Turner, 'Outline of a Theory of Citizenship (originally published in *Sociology* 24 (2): 189–217)', 211.
31. L. Boltanski and E. Chiapello, *The New Spirit of Capitalism*, trans. G. Elliott (London: Verso Books, 2005 [1999]), 22.
32. *n.b.* The terms *la cité, des cités* are translated variously as the city, the polity and political communities. Boltanski and Chiapello, *The New Spirit of Capitalism*, 22, italics in original.
33. Boltanski and Chiapello, *The New Spirit of Capitalism*, 50, n. 54.
34. Ibid., 23.
35. E. Chiapello, 'Reconciling the Two Principal Meanings of the Notion of Ideology: The Example of the Concept of the "New Spirit of Capitalism"', *European Journal of Social Theory* 6, no. 2 (2003), 155–6.
36. L. Boltanski and L. Thévenot, *On Justification: Economies of Worth*, ed. P. DiMaggio et al., trans. C. Porter, *Princeton Studies in Cultural Sociology* (Princeton: Princeton University Press, 2006 [1991]), 374.
37. J.S. Dryzek, *Deliberative Democracy and Beyond: Liberals, Critics, Contestations* (Oxford: Oxford University Press, 2000), 82, 83.
38. Ibid., 83.
39. Gauchet, Renouvin, and Rothnie, 'Democracy and the Human Sciences: An Interview with Marcel Gauchet', 149.
40. Turner, 'Outline of a Theory of Citizenship (originally published in *Sociology* 24 (2): 189–217)', 193–4.
41. Habermas, *Structural Transformation*, 28.
42. Boltanski and Thévenot, *On Justification: Economies of Worth*: 53, 56.
43. Habermas, *Structural Transformation*, 32–3, 35, 56.
44. J.M. Meyer, *Political Nature: Environmentalism and the Interpretation of Western Thought* (Cambridge: MIT Press, 2001), 5.
45. M. Berman, *All That Is Solid Melts Into Air: The Experience of Modernity* (New York: Simon & Schuster, 1982), 345.
46. Lefort, 'Outline of the Genesis of Ideology in Modern Societies', 183–4.
47. Turner, 'Outline of a Theory of Citizenship (originally published in *Sociology* 24 (2): 189–217)', 202; ——, 'Marshall, Social Rights and English Identity', 68.
48. Chiapello, 'Reconciling the Two Principal Meanings of the Notion of Ideology' 166–7, italics in original.
49. A.O. Hirschman, *The Rhetoric of Reaction: Perversity, Futility, Jeopardy* (Cambridge: Belknap Press, 1991); J-W Müller, 'Comprehending Conservatism: A New Framework for Analysis', *Journal of Political Ideologies* 11, no. 3 (2006): 359–65.
50. L. Dumont, 'Sur L'ideologie Politique Française: Une Perspective Comparative', *Le Debat* 58 (1980).

51. S. Lukes, 'Epilogue: The Grand Dichotomy of the Twentieth Century', in *The Cambridge History of Twentieth-Century Political Thought*, ed. T. Ball and R. Bellamy (Cambridge: Cambridge University Press, 2006), 608.
52. N. Bobbio, *Left and Right: The Significance of a Political Distinction*, trans. A. Cameron (Cambridge: Polity, 1996 [1995]), 60–1.
53. Lukes, 'Epilogue: The Grand Dichotomy of the Twentieth Century', 611–12.
54. T. Honderich, *Conservatism: Burke, Nozick, Bush, Blair?*, rev. ed. (London: Pluto Press, 2005 [1989]).
55. L. Thévenot, 'The Plurality of Cognitive Formats and Engagements: Moving between the Familiar and the Public', *European Journal of Social Theory* 10, no. 3 (2007): 411, 414.
56. Habermas, *Structural Transformation*, 134–5.
57. Turner, 'Outline of a Theory of Citizenship (originally published in *Sociology* 24 (2): 189–217)', 193.

3 Challenging Modern Artificialism

1. B.S. Turner, 'Citizenship Studies: A General Theory', *Citizenship Studies* 1, no. 1 (1997): 5–18; B.S. Turner and C. Rojek, *Society and Culture: Principles of Scarcity and Solidarity* (London: Sage, 2001).
2. E.P. Thompson, *The Making of the English Working Class* (London: Penguin Books, 1980 [1963]).
3. T.H. Marshall, 'Citizenship and Social Class', in *The Citizenship Debates: A Reader*, ed. G. Shafir (Minneapolis: University of Minnesota Press, 1998 [1950]), 46.
4. M. Gauchet, 'Democratic Pacification and Civic Desertion', *Thesis Eleven* 29 (1991): 2.
5. N. Fraser and L. Gordon, 'Contract versus Charity: Why Is There No Social Citizenship in the United States', *Socialist Review* 23, no. 3 (1992).
6. Turner, 'Citizenship Studies: A General Theory', 7.
7. J. Valdivielso, 'Social Citizenship and the Environment', *Environmental Politics* 14, no. 2 (2005): 241.
8. B.S. Turner, 'The Erosion of Citizenship', *British Journal of Sociology* 52, no. 2 (2001): 194.
9. A. Giddens, *The Class Structure of the Advanced Societies*, ed. P.S. Cohen, Sociology (London: Hutchinson & Co. Ltd, 1974), 286.
10. See N. Fraser, in N. Fraser and A. Honneth, *Redistribution or Recognition? A Political-Philosophical Exchange*, trans. J. Golb, J. Ingram, and C. Wilke (London: Verso Books, 2003), 80, 106. Fraser refers to a similar point, made by G. Esping-Andersen, *The Three Worlds of Welfare Capitalism* (Princeton: Princeton University Press, 1990).
11. R. Brenner, *The Boom and The Bubble*, 2nd ed. (London: Verso, 2005).
12. L. Sklair, 'Sociology of the Global System', in *The Globalization Reader*, ed. F.J. Lechner and J. Boli (Oxford: Blackwell Publishing, 2005 [2002]), 74.
13. L. Boltanski and L. Thévenot, *On Justification: Economies of Worth*, ed. P. DiMaggio et al., trans. C. Porter, *Princeton Studies in Cultural Sociology* (Princeton: Princeton University Press, 2006 [1991]), 374.

14. J. Lears, *Fables of Abundance* (New York: Harper Collins Publishing, 1994), 345; see also ——, 'Reconsidering Abundance: A Plea for Ambiguity', in *Getting and Spending*, ed. S. Strasser et al. (Washington, DC: German Historical Institute, 1998), 457–8.
15. See R. Behr, 'Anti-Consumerism: New Frontier or Dead End for Progressive Politics?', *Public Policy Research* 17, no. 2 (2010): 124; M. Featherstone, 'Automobilities: Introduction to the Special Edition', *Theory, Culture & Society* 21, no. 4/5 (2004); M. Hilton, 'Consumers and the State Since the Second World War', *ANNALS* 611 (2007); A. Offer, *The Challenge of Affluence: Self-Control and Well-Being in the United States and Britain since 1950* (Oxford: Oxford University Press, 2006); K.T. Jackson, *Crabgrass Frontier: The Suburbanization of the United States* (Oxford: Oxford University Press, 1985); J. Urry, 'The "System" of Automobility', *Theory, Culture & Society* 21, no. 4/5 (2004).
16. M. Featherstone, *Consumer Culture and Postmodernism*, ed. M. Featherstone, 2nd ed., Theory, Culture & Society (London: Sage Publishers, 1996), 35; D. Harvey, *The Condition of Postmodernity: An Enquiry into the Origins of Cultural Change* (Oxford: Blackwell Books, 1990).
17. G. Schulze, 'From Situations to Subjects: Moral Discourse in Transition', in *Constructing the New Consumer Society*, ed. P. Sulkunen et al. (New York: St Martin's Press, 1997), 43.
18. Ibid.; ——, *The Experience Society* (London: Sage Publications, 1995).
19. See J.K. Galbraith, *The Affluent Society* (London: Hamish Hamilton, 1958).
20. J. Baudrillard, *For a Critique of the Political Economy of the Sign*, trans. C Levin (St. Louis: Telos Press, 1981 [1972]), 29–33, 88–90.
21. Schulze, 'From Situations to Subjects: Moral Discourse in Transition', 39, 41, 43, 48–9.
22. I. Blühdorn, *Post-ecologist Politics: Social theory and the Abdication of the Ecologist Paradigm* (London: Routledge, 2000); R. Inglehart, *The Silent Revolution: Changing Values and Political Styles Among Western Publics* (Princeton: Princeton University Press, 1977); ——, *Culture Shift in Advanced Industrial Societies* (Princeton: Princeton University Press, 1990); ——, *Modernization and Postmodernization* (Princeton: Princeton University Press, 1997); T. Roszak, *The Making of a Counterculture: Reflections on the Technocratic Society & Its Youthful Opposition*, 2nd ed. (Berkeley: University of California Press, 1995 [1968]); C. Lasch, *Haven in a Heartless World: The Family Besieged* (New York: Basic Books, 1977); ——, *The Culture of Narcissism: American Life in An Age of Diminishing Expectations* (New York: W.W. Norton and Co., 1979).
23. See R. Dunlap and G.A. Mertig, 'Global Concern for the Environment: Is Affluence a Prerequisite?', *Journal of Social Issues* 51, no. 3 (1995); ——, 'Global Environmental Concern: An Anomaly for Postmaterialism', *Social Science Quarterly* 78, no. 1 (1997); J. Martinez-Alier, *The Environmentalism of the Poor: A Study of Ecological Conflicts and Valuation* (Cheltenham: Edward Elgar, 2002).
24. D. Bell, *The End of Ideology* (New York: Glencoe Free Press, 1960); ——, *The Coming of Postindustrial Society* (New York: Basic Books, 1973); ——, *The Cultural Contradictions of Capitalism*, 2nd ed. (London: Heinemann Books, 1978), 198.
25. Ibid., 201–2.

26. Ibid., n. 198.
27. Ibid., 283, italics in original.
28. M. Crozier, S.P. Huntington, and J. Watanuki, *The Crisis of Democracy: Report on the Governability of Democracies to the Trilateral Commission* (New York: NYU Press, 1975); I. Kristol 'Keeping Up with Ourselves', in *The End of Ideology*, ed. C.I. Waxman (New York: Funk & Wagnalls, [2006]1968).
29. Indeed, Bell states in his Preface, 'religion…is the fulcrum of the book' and poses the question, 'Can we set a limit to *hubris*?' Bell, *The Cultural Contradictions of Capitalism*, xxix.
30. Giddens, *The Class Structure of the Advanced Societies*, 292–3.
31. J. Habermas, *Legitimation Crisis*, trans. T. McCarthy (Boston Beacon Press, 1975 [1973]), 123–4, Italics in original. See also, P. Bachrach, *The Theory of Democratic Elitism* (London: University of London Press, Ltd., 1967), 5–6, 8–9, 103.
32. J. Habermas, *Toward a Rational Society: Student Protest, Science, and Politics*, trans. J.J. Shapiro (London: Heinemann, 1971 [1968]), 110, italics in original.
33. See, for example, Giddens, *The Class Structure of the Advanced Societies*, 61–3, 85, 257. Z. Bauman, *Work, Consumerism and the New Poor* (Buckingham: Open University Press, 1998); ——, 'The Poor – and the Rest of Us', *Arena* New Series, no. 12 (1998); M. Peel, *The Lowest Rung: Voices of Australia's Poverty* (Cambridge: Cambridge University Press, 2004), Z. Bauman, 'Social Issues of Law and Order', *British Journal of Criminology* 40 (2000): 207. R. Brenner, 'New Boom or New Bubble?', *New Left Review* II, no. 25 (2004); G. Duménil and D. Lévy, 'Neoliberal Income Trends', *New Left Review* II, no. 30 (2004); Harvey, *The Condition of Postmodernity: An Enquiry into the Origins of Cultural Change*, 330–5; K. Eder, *The New Politics of Class: Social Movements and Cultural Dynamics in Advanced Societies* (London: Sage, 1993).
34. J.S. Dryzek, *Deliberative Democracy and Beyond: Liberals, Critics, Contestations* (Oxford: Oxford University Press, 2000), 95.
35. Habermas, *Legitimation Crisis*; C. Offe, *Contradictions of the Welfare State* (Cambridge: MIT Press, 1984); J. O'Connor, *The Fiscal Crisis of the State* (New York: Transaction Books, 2009 [1973]); ——, *Accumulation Crisis* (Oxford: Basil Blackwell, 1984).
36. J. Heath, *Efficient Society: Why Canada Is as Close to Utopia as It Gets* (Toronto: Penguin Global, 2005); J. Heath and A. Potter, *Rebel Sell: Why the Culture Can't Be Jammed* (New York: HarperBusiness, 2004).
37. Habermas, *Toward a Rational Society: Student Protest, Science, and Politics*, 121–2; ——, *Lifeworld and System: A Critique of Functionalist Reason*, trans. T. McCarthy, The Theory of Communicative Action: Vol. 2 (Boston Beacon Press, 1987 [1985]), 388.
38. P. Rosanvallon, *The New Social Question: Rethinking the Welfare State* (Princeton: Princeton University Press, 2000 [1995]), 49.

4 The New Citizenship, Imperatives of State and Questions of Justice

1. M. Davis, 'The Political Economy of Late Imperial America', *New Left Review* 143 (1984): 33–4.

2. J. Habermas, *Lifeworld and System: A Critique of Functionalist Reason*, trans. T. McCarthy, The Theory of Communicative Action: Vol. 2 (Boston Beacon Press, 1987 [1985]), 388.
3. T. Eagleton, *The Illusions of Postmodernism* (London: Blackwell Publishing, 1996).
4. L. Dumont, *Essays on Individualism: Modern Ideology in Anthropological Perspective*, English ed. (Chicago: University of Chicago Press, 1986 [1983]), 217.
5. J. Habermas, *Toward a Rational Society: Student Protest, Science, and Politics*, trans. J.J. Shapiro (London: Heinemann, 1971 [1968]), 75, 122; R. Debray, 'A Modest Contribution to the Rites and Ceremonies of the Tenth Anniversary', *New Left Review* I, no. 115 (1979): 53–4.
6. See, for example, G.R. Krippner, 'The Financialization of the American Economy', *Socio-Economic Review*, no. 3 (2005); G. Duménil and D. Lévy, *Capital Resurgent: Roots of the Neoliberal Revolution*, trans. D. Jeffers (Cambridge Harvard University Press, 2004).
7. See, for example, D. Harvey, *The Condition of Postmodernity: An Enquiry into the Origins of Cultural Change* (Oxford: Blackwell Books, 1990); S. Sassen, *Globalization and Its Discontents: Essays on the New Mobility of People and Money* (New York: New Press, 1998); P. Dicken, *Global Shift: Transforming the World Economy*, 3rd ed. (London: Sage Books, 1999).
8. L. Boltanski and E. Chiapello, *The New Spirit of Capitalism*, trans. G. Elliott (London: Verso Books, 2005 [1999]), 29.
9. P. Heelas, 'Work Ethics, Soft Capitalism and the "Turn to Life"', in *Cultural Economy*, ed. P. du Gay and M. Pryke (London: Sage Publications, 2002), 68, 78, 180–3.
10. Two examples from the management literature of the early 2000s suffice to demonstrate this tendency. See R.P. Hill and D.L. Stephens, 'The Compassionate Organisation in the 21st Century', *Organizational Dynamics* 32, no. 4 (2003); C. Juniper and M. Moore, 'Synergies and Best Practices of Corporate Partnerships for Sustainability', *Corporate Environmental Strategy* 9, no. 3 (2002).
11. J.B. Schor, *The Overworked American: The Unexpected Decline of Leisure* (New York: Basic Books/HarperCollins Publishing, 1991); B. Ehrenreich, *Nickeld and Dimed: On (not) Getting by in America* (New York: Metropolitan Press, 2002); ——, *Bait and Switch: The Futile Pursuit of the Corporate Dream* (London: Granta, 2006); M. Bunting, *Willing Slaves: How the Overwork Culture Is Ruling Our Lives* (London: HarperCollins Publishers, 2004); A. Hochschild, *The Time Bind: When Work Becomes Home and Home Becomes Work* (New York: Metropolitan Books, 1997).
12. N. Fraser, 'Feminism, Capitalism and the Cunning of History', *New Left Review* II, no. 56 (2009): 110–11.
13. Boltanski and Chiapello, *The New Spirit of Capitalism*, 419–20.
14. Z. Bauman, *The Individualized Society* (Cambridge: Polity Press, 2001), 75–7; ——, *Consuming Life* (London: Polity Press, 2007).
15. T. Frank, *The Conquest of Cool: Business Culture, Counterculture, and the Rise of Hip Consumerism* (Chicago: University of Chicago Press, 1997); G. Ritzer, *Enchanting a Disenchanting World: Revolutionising the Means of Consumption* (Thousand Oaks: Pine Forge Press, 1999); C. Campbell, *The Romantic Ethic*

and the Spirit of Consumerism (Oxford: Blackwell, 1987); A. Scerri, 'Triple Bottom-line Capitalism and the 3rd Place', *Arena Journal New Series*, no. 20 (2003); A. Ross, *No Collar: The Humane Workplace and Its Hidden Costs* (New York: Basic Books, 2002).

16. Boltanski and Chiapello, *The New Spirit of Capitalism*, 356.
17. A. Honneth, 'Organized Self-Realization: Some Paradoxes of Individualization', *European Journal of Social Theory* 7, no. 4 (2004): 474.
18. Boltanski and Chiapello, *The New Spirit of Capitalism*, 345–6.
19. A. Elliott and C. Lemert, *The New Individualism: The Emotional Costs of Globalization* (London: Routledge, 2006); R. Sennett, *The Culture of the New Capitalism* (London: Yale University Press, 2005); G. Delanty, *Modernity and Postmodernity: Knowledge, Power, and the Self* (London: Sage Publications, 2000), 161.
20. A. Melucci, *The Playing Self: Person and Meaning in the Planetary Society*, ed. J.C. Alexander and S. Seidman (Cambridge: Cambridge University Press, 1996), 36; U. Beck and E. Beck-Gernsheim, *Individualization*, ed. M. Featherstone, trans. P. Camiller (London: Sage Publications, 2002); A. Giddens, *Modernity & Self Identity* (Cambridge: Polity Press, 1991); H. Joas, *The Creativity of Action*, trans. J. Gaines and P. Keast (Chicago: University of Chicago Press, 1996 [1992]); C. Taylor, *Sources of the Self: The Making of Modern Identity* (Cambridge: Harvard University Press, 1989); ——, *The Ethics of Authenticity* (Cambridge: Harvard University Press, 1991).
21. M. Gauchet, B. Renouvin, and S. Rothnie, 'Democracy and the Human Sciences: An Interview with Marcel Gauchet', *Thesis Eleven* 26 (1990 [1988]): 144, 146, italics removed;
22. M. Gauchet, 'A New Age of Personality', *Thesis Eleven* 60 (2000): 34.
23. M. Gauchet, *The Disenchantment of the World: A Political History of Religion*, ed. T. Pavel and M. Lilla, trans. O. Burge, New French Thought (Princeton: Princeton University Press, 1997 [1985]).
24. N. Fraser, *Justice Interruptus: Critical Reflections on the Postsocialist Condition* (London: Routledge, 1997).
25. Beck and Beck-Gernsheim, *Individualization*, 29.
26. U. Beck and E. Beck-Gernsheim, 'Freedom's Children', trans. P. Camiller. Chap. 12 in *Individualization*, ed. U. Beck and E. Beck-Gernsheim (London: Sage, 2001), 160.
27. Ibid., 162.
28. Ibid.
29. M. Gauchet, 'Democratic Pacification and Civic Desertion', *Thesis Eleven* 29 (1991): 7.
30. N. Rose, *Powers of Freedom: Reframing Political Thought* (Cambridge: Cambridge University Press, 1999); P. Miller and N. Rose, *Governing the Present: Administering Economic, Social and Personal Life* (Cambridge: Polity, 2008).
31. See, for example, J.A. Summerville, B.A. Adkins, and G. Kendall, 'Community Participation, Rights, and Responsibilities: The Governmentality of Sustainable Development Policy in Australia', *Environment and Planning C: Government and Policy* 26 (2008); M. Taylor, 'Community Participation in the Real World: Opportunities and Pitfalls in New Governance Spaces', *Urban Studies* 44, no. 2 (2009).

32. M. Gunder, 'Sustainability: Planning's Saving Grace or Road to Perdition?', *Planning Education and Research* 26 (2006): 209, italics in original.
33. M. Raco, 'Securing Sustainable Communities: Citizenship, Safety and Sustainability in the New Urban Planning', *European Urban and Regional Studies* 14, no. 4 (2007): 109.
34. N. Brenner, *Spaces of Neoliberalism: Urban Restructuring in North America and Western Europe* (London: Wiley-Blackwell, 2003); B. Jessop, 'The Changing Governance of Welfare: Recent Trends in Its Primary Functions, Scale and Modes of Coordination', *Social Policy and Administration* 33, no. 4 (1999); ——, 'Liberalism, Neoliberalism and Urban Governance: A State-Theoretical Perspective', *Antipode* 34, no. 3 (2002); R. Keil, '"Common-Sense" Neoliberalism: Progressive Conservative Urbanism in Toronto, Canada', *Antipode* 34, no. 3 (2002).
35. J.N. Rosenau, 'Governance, Order and Change in World Politics', in *Governance without Government*, ed. J.N. Rosenau and E-O. Czemoiel (Cambridge: Cambridge University Press, 1992); R.A.W. Rhodes, *Understanding Governance* (Buckingham: Open University Press, 1997); B. Eggers, *Government by Network* (Washington: Brookings Institute Press, 2005); J. Pierre, ed., *Debating Governance* (Oxford: Oxford University Press, 2000); J. Pierre and B.G. Peters, *Governance, Policy and the State* (London: Macmillan, 2000); M. Hajer and H. Wagenaar, eds., *Deliberative Policy Analysis: Understanding Governance in the Network Society* (Cambridge: Cambridge University Press, 2003).
36. E. Swyngedouw, 'Impossible "Sustainability" and the Postpolitical Condition', in *The Sustainable Development Paradox: Urban Political Economy in the United States and Europe*, ed. R. Kreuger and D. Gibbs (New York: The Guilford Press, 2007), 25–6, 33, 38.
37. K. Hobson, 'Thinking Habits into Action: The Role of Knowledge and Process in Questioning Household Consumption Practices', *Local Environment* 8, no. 1 (2003): 95–112; G. Seyfang and A. Smith 'Grassroots Innovation for Sustainable Development: Towards a New Research and Policy Agenda', *Environmental Politics* 16, no. 4 (August 2007): 584–603; Raco, 'Securing Sustainable Communities: Citizenship, Safety and Sustainability in the New Urban Planning'.
38. M. Hajer, 'Policy without Polity? Policy Analysis and the Institutional Void', *Policy Sciences* 36 (2003): 175, 177.
39. A. Jordan, R. Wurzel, and A. Zito, 'The Rise of "New" Policy Instruments in Comparative Perspective: Has Governance Eclipsed Government?', *Political Studies* 53, no. 3 (2005): 477–8.
40. D. Torgerson, 'Democracy through Policy Discourse', in *Deliberative Policy Analysis: Understanding Governance in the Network Society*, ed. M. Hajer and H. Wagenaar (Cambridge: Cambridge University Press, 2003).
41. L. Elliott, 'Global Environmental Governance', in *Global Governance: Critical Perspectives*, ed. R. Wilkinson and S. Hughes (London: Routledge, 2002), 71.
42. P. Wagner, 'After *Justification:* Repertoires of Justification and the Sociology of Modernity', *European Journal of Social Theory* 2, no. 3 (1999); B.S. Turner, 'Extended Review: Justification, the City, and Late Capitalism', *The Sociological Review* 55, no. 2 (2007); T.M. Kemple, 'Spirits of Late Capitalism', *Theory, Culture & Society* 24, no. 3 (2007).

43. Melucci, *The Playing Self: Person and Meaning in the Planetary Society*; M. Maffesoli, *The Time of the Tribes: The Decline of Individualism in Mass Society*, ed. M. Featherstone, Theory, Culture & Society (London: Sage Books, 1996 [1988]); ——, 'The Return of Dionysus', in *Constructing the New Consumer Society*, ed. P. Sulkunen et al. (New York: St Martin's Press, 1997).
44. C. Lafaye, and L. Thévenot 'Une Justification Écologique?: Conflits Dans L'aménagement De La Nature', *Revue Française de Sociologie* 34, no. 4 (1993): 495–524; L. Thévenot, M. Moody, and C. Lafaye, 'Forms of Valuing Nature: Arguments and Modes of Justification in French and American Environmental Disputes', in *Rethinking Comparative Cultural Sociology*, ed. L. Thévenot and M. Lamont (Cambridge: Cambridge University Press, 2001); M. Moody and L. Thévenot 'Comparing Models of Strategy, Interests and the Public Good in French and American Environmental Disputes', in *Rethinking Comparative Cultural Sociology*, ed. L. Thévenot and M. Lamont (Cambridge: Cambridge University Press, 2001).
45. R. Scruton, 'Conservatism', in *Political Theory and the Ecological Challenge*, ed. A. Dobson and R. Eckersley (Cambridge: Cambridge University Press, 2006).
46. Boltanski and Chiapello, *The New Spirit of Capitalism*, 354.
47. Ibid., 348.
48. M. Castells, 'Materials for an Exploratory Theory of the Network Society', *British Journal of Sociology* 51, no. 1 (2000): 19.
49. K. Brown, 'Human Development and Environmental Governance: a Reality Check', in *Governing Sustainability*, ed. W.N. Adger and A. Jordan (Cambridge: Cambridge University Press, 2009), 42, 47.
50. A. Weale, 'Governance, Government and the Pursuit of Sustainability', in *Governing Sustainability*, ed. W.N. Adger and A. Jordan (Cambridge: Cambridge University Press, 2009), 55.
51. J.S. Dryzek, *Deliberative Global Politics* (Cambridge: Polity, 2006), 107.
52. J. Gupta, 'North-South Aspects of the Climate Change Issue: Towards a Negotiating Theory and Strategy for Developing Countries', *International Journal of Sustainable Development* 3, no. 2 (2000), 119; L. Lohmann, *Carbon Trading: A Critical Conversation on Climate Change, Privatisation and Power*, ed. N. Hällström, O. Nordberg, and R. Österbergh, vol. 48, Development Dialogue (Uppsala: Dag Hammarskjöld Centre/Corner House, 2006), 62.
53. G. Gereffi, R. Garcia-Johnson, and E. Sasser, 'The NGO-Industrial Complex', *Foreign Policy* 125 (2001); L.K. Petersen, 'Changing Public Discourse on the Environment: Danish Media Coverage of the Rio and Johannesburg un Summits', *Environmental Politics* 16, no. 2 (2007).
54. Dryzek, *Deliberative Global Politics*, 155.
55. S.B. Banerjee, 'Corporate Social Responsibility: The Good, the Bad and the Ugly', *Critical Sociology* 34, no. 1 (2008): 51; A. Giddens, 'The Politics of Climate Change: National Responses to the Challenge of Global Warming' (London: Policy Network, 2008), 10.
56. P. Wagner, *Modernity as Experience and Interpretation: A New Sociology of Modernity* (Cambridge: Polity Press, 2008).
57. Lafaye and Thévenot, 'Une Justification Écologique?: Conflits Dans L'aménagement De La Nature'; J.S. Dryzek, *Deliberative Democracy and Beyond: Liberals, Critics, Contestations* (Oxford: Oxford University Press, 2000), 98.

58. J. Connelly, 'The Virtues of Environmental Citizenship', in *Environmental Citizenship*, ed. A. Dobson and D. Bell (Cambridge: MIT Press, 2006).
59. S. Dryzek and P. Dunleavy, *Theories of the Democratic State* (New York: Palgrave Macmillan, 2009), 253–4.

5 Not Just the Warm, Fuzzy Feeling You Get from Buying Free-Range Eggs ...

1. See, for example, S. Jenkins, *Thatcher and Sons: A Revolution in Three Acts* (London: Penguin Books, 2006); D. Sassoon, *One Hundred Years of Socialism: The West European Left in the Twentieth Century* (London: I.B. Tauris, 2010); J. Olsen, M. Koss, and D. Hough, 'From Pariahs to Players? Left Parties in National Governments', in *Left Parties in National Governments*, ed. J. Olsen, M. Koss, and D. Hough (London: Palgrave Macmillan, 2010); T. Frank, *What's the Matter with Kansas?* (New York: Vintage Books, 2004); J.C. Alexander, *The Performance of Politics: Obama's Victory and the Democratic Struggle for Power* (Oxford: Oxford University Press, 2010); C. Hamilton, 'What's Left? The Death of Social Democracy', *Quarterly Essay*, no. 21 (2006).
2. See S. Buckler and D.P. Dolowitz, 'Theorizing the Third Way: New Labour and Social Justice', *Journal of Political Ideologies* 5, no. 3 (2000); H. Dean, 'Popular Discourse and the Ethical Deficiency of "Third Way" Conceptions of Citizenship', *Citizenship Studies* 8, no. 1 (2004); A. Giddens, *The Third Way* (Cambridge: Polity Press, 1998); J. Clarke, 'New Labour's Citizens: Activated, Empowered, Responsibilized, Abandoned?', *Critical Social Policy* 25, no. 4 (2005); B. Burkitt and F. Ashton, 'The Birth of the Stakeholder Society', *Critical Social Policy* 16, no. 49 (1996); A. Giddens, 'The Politics of Climate Change: National Responses to the Challenge of Global Warming' (London: Policy Network, 2008); ——, *The Politics of Climate Change* (Cambridge: Polity Press, 2009).
3. A. Callinicos, *Equality*, Themes for the 21st Century (Cambridge: Polity Press, 2000), 41.
4. R. Prabhakar, 'Stakeholding: Does It Possess a Stable Core?', *Journal of Political Ideologies* 8, no. 3 (2003): 347.
5. Callinicos, *Equality*, 36.
6. T. Turner, 'Shifting the Frame from Nation-State to Global Market', *Social Analysis* 46, no. 2 (2002): 56, 68–9.
7. G. Kelly, D. Kelly, and A. Gamble. *Stakeholder Capitalism* (London: Macmillan, 1997), 44, 228.
8. G. Duncan, 'After Happiness' *Journal of Political Ideologies* 12, no. 1 (2007); Jenkins, *Thatcher and Sons: A Revolution in Three Acts*; M. Glasman, 'Society not State: The Challenge of the Big Society', *Public Policy Research* 17, no. 2 (2010); R.S. Grayson, 'Localism the American Way', *Public Policy Research* 17, no. 2 (2010); E. Mayo, 'Dreams of a Social Economy', *Public Policy Research* 17, no. 2 (2010); G. Smith, 'Green Citizenship and the Social Economy', *Environmental Politics* 14, no. 2 (2005); Buckler and Dolowitz, 'Theorizing the Third Way: New Labour and Social Justice'; Dean, 'Popular Discourse and the Ethical Deficiency of "Third Way" Conceptions of Citizenship'; Giddens, *The Third Way*; ——, *The Third Way and Its Critics* (Cambridge:

Cambridge University Press, 2000); C. Pierson, 'Lost property: What the Third Way Lacks', *Journal of Political Ideologies* 10, no. 2 (2005); D. Smith, 'Social Exclusion, the Third Way and the Reserve Army of Labour', *Sociology* 41, no. 2 (2007); J. Handler, 'Reforming/Deforming Welfare', *New Left Review* II, no. 4 (2000); T. Honderich, *Conservatism: Burke, Nozick, Bush, Blair?*, Rev. ed. (London: Pluto Press, 2005 [1989]).

9. A. Supiot, 'Law and Labour: A World Market of Norms?', *New Left Review* I, no. 39 (2006): 109, 18.
10. M. Scott, 'Reflections on the Big Society', *Community Development Journal* 46, no. 1 (2010).
11. N. Boles, *Which Way's Up: The Future for Coalition Britain and How to Get There* (London: Biteback Publishers, 2010), 9; Prime Minister's Office, 'Government Launches 'Big Society' Programme', Office of the Prime Minister, http://www.number10.gov.uk/news/topstorynews/2010/05/big-society-50248; P. Totaro, 'Britain Surprises with Carbon Plan', *The Age*, May 21 2011.
12. D. Roberts, 'The Full Text of Obama's Energy Remarks', Grist Magazine inc., http://www.grist.org/article/obama_factsheet; B. Obama, 'Inaugural Address' (2009), http://obamaspeeches.com. Of course, 'ecological modernization' at the Federal scale in the United States remains a problematic discourse. However, I contend that even though the situation surrounding domestic *realpolitik* within the US keeps hopes for greening the economy off the agenda, the primary mode of appeal to citizens by both major parties remains one of providing security from all manner of 'risks'.
13. L. Dumont, *German Ideology: From France to Germany and Back* (Chicago: Chicago University Press, 1994 [1991]), 9.
14. P. Alston, ed., *Labour Rights as Human Rights* (Oxford: Oxford University Press, 2005); S. Deakin and F. Wilkinson, *The Law of the Market: Industrialization, Employment, and Legal Evolution* (Oxford: Oxford University Press, 2005); M. Freedland, *The Personal Employment Contract* (Oxford: Oxford University Press, 2003). P. Dwyer, 'Conditional Citizens? Welfare Rights and Responsibilities in the Late 1990s', *Critical Social Policy* 18, no. 4 (1998); Handler, 'Reforming/Deforming Welfare'; P. Harris, 'From Relief to Mutual Obligation: Welfare Rationalities and Unemployment in 20th-Century Australia', *Journal of Sociology* 37, no. 1 (2001); Jessop, 'The Changing Governance of Welfare: Recent Trends In Its Primary Functions, Scale and Modes of Coordination'; R. Sainsbury, '21st Century Welfare – Getting Closer to Radial Benefit Reform', *Public Policy Research* 17, no. 2 (2010).
15. P. Rosanvallon, *The New Social Question: Rethinking the Welfare State* (Princeton: Princeton University Press, 2000 [1995]), 211.
16. B. Zimmerman, 'Pragmatism and the Capability Approach: Challenges in Social Theory and Empirical Research', *European Journal of Social Theory* 9, no. 4 (2006): 468.
17. van Steenbergen, 'Towards a Global Ecological Citizen', 47, 143–4; B.S. Turner, 'The Erosion of Citizenship', *British Journal of Sociology* 52, no. 2 (2001): 199.
18. J. Martinez-Alier, *The Environmentalism of the Poor: A Study of Ecological Conflicts and Valuation* (Cheltenham: Edward Elgar, 2002), 5–10.
19. J. Barry, 'Ecological Modernization', in *Debating the Earth*, ed. J.S. Dryzsek and D. Schlosberg (Oxford: Oxford University Press, 2005), 311.

20. J.S. Dryzek, *Deliberative Democracy and Beyond: Liberals, Critics, Contestations* (Oxford: Oxford University Press, 2000), 143; ——, *Politics of the Earth*, 2nd ed. (Oxford: Oxford University Press, 2004), 167; also, J.S. Dryzek et al., *Green States and Social Movements: Environmentalism in the United States, United Kingdom, Germany and Norway* (Oxford: Oxford University Press, 2003).
21. D. Schlosberg and S. Rinfret, 'Ecological Modernisation, American Style', *Environmental Politics* 17, no. 2 (2008), 265; see also J. Barry and P. Doran, 'Refining Green Political Economy: From Ecological Modernisation to Economic Security and Sufficiency', *Analyse & Kritik* 28, no. 2 (2006).
22. For Ted Honderich, the justification of selfishness has always been the 'essence' of conservatism. My point is that there is less pressure to defend this 'essence' in terms of upholding tradition, favouring particularism over universalism and essentialism over contingency in the context of stakeholder citizenship. See Honderich, *Conservatism: Burke, Nozick, Bush, Blair?*
23. J. Elkington, *Cannibals with Forks: The Triple Bottom-Line of 21st Century Business* (Gabriola Island: New Society Publishers, 1997); ——, *The Chrysalis Economy* (Oxford: Capstone Press, 2001).
24. See http://www.accountability.org/services/stakeholder.html; http://www.globalreporting.org/ReportingFramework/G3Guidelines/; http://www.unglobalcompact.org/ParticipantsAndStakeholders/index.html International Standards Organization, 'Guidance on Social Responsibility', ISO. http://www.iso.org/iso/about.htm.
25. See United Nations Global Compact, 'United Nations Global Compact: About the GC', UNGC, www.unglobalcompact.org; Global Reporting Initiative, 'Global Reporting Initiative: Home', Global Reporting Initiative, www.globalreporting.org.
26. A. Salmon, *La Tentation Éthique Du Capitalisme* (Paris: La Découverte 2007); ——, *Moraliser Le Capitalisme* (Paris: CNRS Éditions, 2009).
27. Fiona Haddock, 'Socially Responsible Companies: Corporate Angels', *Global Finance* 13, no. 12 (1999): 24, 25.
28. G. Atkinson, 'Measuring Corporate Sustainability', *Journal of Environmental Planning and Management* 43, no. 2 (2000): 237.
29. Gilding, P. (2002) *Single Bottom Line Sustainability*, The Sydney Papers, no volume given: 53–60.
30. See, for example, A. Lagan, 'a Revolution to Find Value in Common Ground', *Australian Financial Review*, 29 May 2003, 23; S. Zuboff, 'Evolving: A Call to Action', *Fast Company* Social Capitalists Issue, no. 78 (2004): 97; R. Kelly, 'Community Values Create Value', *Business Review Weekly*, 16 January 2003, 447; C. Juniper and M. Moore, 'Synergies and Best Practices of Corporate Partnerships for Sustainability', *Corporate Environmental Strategy* 9, no. 3 (2002), 267–9; R.P. Hill and D.L. Stephens, 'The Compassionate Organisation in the 21st Century', *Organizational Dynamics* 32, no. 4 (2003); 331–41; M. Benioff and K. Southwick, *Compassionate Capitalism: How Corporations Can Make Doing Good an Integral Part of Doing Well* (New York: Career Press, 2003).
31. KPMG, 'Beyond Numbers™: How Leading Organisations Link Values with Value to Gain Competitive Advantage', KPMG, www.kpmg.com.
32. R. Hill, 'Introducing the Triple Bottom-line', KPMG Global, http://www.kpmg.com/Global/en/IssuesAndInsights/ArticlesPublications/Press-releases/Pages/Press-release-Introducing-the-triple-bottom-line-1-Mar-2010.aspx.

33. KPMG Climate Change & Sustainability Services, 'KPMG Global: KPMG Climate Change & Sustainability Services', KPMG Global, http://www.kpmg.com/Global/en/WhatWeDo/Advisory/Risk-Compliance/Internal-Audit/Climate-Change-Sustainability-Services/Pages/Climate-change-Sustainability-Services.aspx.
34. Martinez-Alier, *The Environmentalism of the Poor: A Study of Ecological Conflicts and Valuation*, 29.
35. W. Lazonick and M. O'Sullivan, 'Maximizing Shareholder Value: A New Ideology for Corporate Governance', *Economy and Society* 29, no. 1 (2000); C. Stoney and D. Winstanley, 'Stakeholding: Confusion or Utopia? Mapping the Conceptual Terrain', *Journal of Management Studies* 38, no. 5 (2001): 608.
36. I. Porter, 'General Electric Warms Up a Slice Nuclear Pie', *The Age*, 26 May 2006, 2. General Electric Company, 'Ongoing Stakeholder Engagement', GE, http://www.ge.com/citizenship/programs-activities/business-processes/ongoing-stakeholder-engagement.html.
37. See A. Scerri, 'Paradoxes of Increased Individuation and Public Awareness of Environmental Issues', *Environmental Politics* 18, no. 4 (2009): 476.
38. T. Gray-Barkan, 'Wal-Mart and Beyond: The Battle for Good Jobs and Strong Communities in Urban America' (Los Angeles: Los Angeles Alliance for a New Economy/Partnership for Working Families, 2007); L. Sandercock, 'The 'Wal-Marting' of the World? The Neoliberal City in North America', *City & Community* 9, no. 1 (2005).
39. T.P. Lyon and J.W. Maxwell, 'Corporate Social Responsibility and the Environment: A Theoretical Perspective', *Review of Environmental Economics and Policy* 2, no. 2 (2008): 252.
40. P.R. Portney, 'The (Not So) New Corporate Social Responsibility: An Empirical Perspective', *Review of Environmental Economics and Policy* 2, no. 2 (2008): 262, italics in original.
41. H.S. Brown, M. de Jong, and T. Lessidrenska, 'The Rise of the Global Reporting Initiative: A Case of Institutional Entrepreneurship', *Environmental Politics* 18, no. 2 (2009).
42. UNEP GRI, KPMG, UCGA, 'Carrots and Sticks – Promoting Transparency and Sustainability: An Update on Trends in Mandatory and Voluntary Sustainability Reporting' (Nairobi: Global Reporting Initiative, United Nations Environment Programme, KPMG, Unit for Corporate Governance in Africa, 2009), 8.
43. See M.L. Parry et al., eds., *Fourth Assessment Report of the Intergovernmental Panel on Climate Change* (Cambridge: Cambridge University Press, 2007).
44. A.P.J. Mol and G. Spaargaren, 'Environment, Modernity and the Risk-Society', *International Sociology* 8, no. 4 (1993): 357.
45. M. Rayner, R. Harrison, and S. Irving, 'Ethical Consumerism: Democracy through the Wallet', *Journal of Research for Consumers* 1, no. 3 (2002); K. Soper, 'Rethinking the "Good Life": The Consumer as Citizen', *Capitalism, Nature, Socialism* 15, no. 3 (2004); ——, 'Alternative Hedonism, Cultural Theory and the Role of Aesthetic Revisioning', *Cultural Studies* 22, no. 5 (2008).
46. I. Grayson, 'Green Consumerism', *Chain Reaction*, Autumn 1989, 27.
47. M. Schudson, 'The Troubling Equivalence of Citizen and Consumer', *Annals of the American Academy of Political and Social Sciences* 608 (2006): 193–204.

48. G. Spaargaren, and A.P.J. Mol, 'Greening Global Consumption: Redefining Politics and Authority', *Global Environmental Change* 18, no. 2 (2008): 358, emphasis added.
49. E.F. Isin, 'The Neurotic Citizen', *Citizenship Studies* 8, no. 3 (2004): 226; Mol and Spaargaren, 'Environment, Modernity and the Risk-Society' 443–4.
50. Supiot, 'Law and Labour: A World Market of Norms?', 109.
51. Scerri, 'Paradoxes of Increased Individuation and Public Awareness of Environmental Issues', 474, italics in original.
52. K. Hobson, 'Bins, Bulbs and Shower Timers: On the "Techno-Ethics" of Sustainable Living', *Ethics, Place & Environment* 9, no. 3 (2006): 321; ——, 'Thinking Habits into Action: The Role of Knowledge and Process in Questioning Household Consumption Practices' *Local Environment* 8, no. 1 (2003): 95–112, 110.
53. V. Timmer and N-K Seymoar, 'Vancouver Working Group Discussion Paper', in *The World Urban Forum 2006* (Vancouver: UN Habitat – International Centre for Sustainable Cities, 2005); National Endowment for the Arts, *The NEA 1965–2000: A Brief Chronology of Federal Support for the Arts* (Washington DC: NEA, 2000).
54. R. Veenhoven, 'Happiness as an Indicator in Social Policy Evaluation: Some Objections Considered', in *Between Sociology and Sociological Practice*, ed. K. Mesman Schultz et al. (Nijmegen: Institute for Applied Social Sciences, 1993); R. Veenhoven and J. Ehrhardt, 'The Cross-National Pattern of Happiness: Test of Predictions Implied in Three Theories of Happiness', *Social Indicators Research* 34, no. 1 (1995).
55. Marsh Mercer Kroll, 'Quality of Living – Survey Highlights', Marsh Mercer Kroll, www.mercer.com/qualityofliving; Economist Intelligence Unit, 'Liveability Ranking', The Economist, http://www.economist.com/markets/rankings/displaystory.
56. J. Peck, 'Struggling with the Creative Class', *International Journal of Urban and Regional Research* 29, no. 4 (2005).
57. F. Fischer, *Democracy & Expertise: Reorienting Policy Inquiry* (Oxford: Oxford University Press, 2009).
58. See, for example, N. Casey, 'Melbourne Benchmarking and Liveability' (Melbourne: City of Melbourne, 2009); ——, 'Melbourne City Research: International City Comparisons' (Melbourne: City of Melbourne, 2011).
59. Veenhoven and Ehrhardt, 'The Cross-National Pattern of Happiness: Test of Predictions Implied in Three Theories of Happiness', 34–6.
60. M.R. Hagerty, 'Unifying Livability and Comparison Theory: Cross-National Time-Series Analysis of Life-Satisfaction', *Social Indicators Research* 47, no. 1 (1999).
61. A. Desrosières, 'Historiciser L'action Publique: L'Etat, Le Marche Et Les Statistiques', in *Historicites de l'action publique*, ed. P. Laborier and D. Trom (Paris: Presses Universitaires de France, 2003); P. Le Galès and A. Scott, 'Une Revolution Bureaucratique Britannique? Autonomie Sans Controle Ou "Freer Markets, More Rules"', *Revue Française de Sociologie* 49, no. 2 (2008); W. Espeland and M. Sauder, 'Rankings and Reactivity: How Public Measures Recreate Social Worlds', *American Journal of Sociology* 113, no. 1 (2007).
62. UNDP, 'Human Development Reports: History of the Human Development Reports', United Nations Development Programme, http://hdr.undp.org/en/

humandev/reports/; OECD, 'Better Life Index', Organization for Economic Co-operation and Development, http://www.oecdbetterlifeindex.org/.

6 Justice after Dualism

1. UNEP, 'Global Green New Deal' (Geneva: United Nations Environment Programme, 2009); ——, 'Towards a Green Economy: Pathways to Sustainable Development and Poverty Eradication' (Geneva: United Nations Environment Programme, 2011); TEEB, 'The Economics of Ecosystems and Biodiversity: Mainstreaming the Economics of Nature: A Synthesis of the approach, conclusions and recommendations of TEEB' (Geneva: The Economics of Ecosystems and Biodiversity/United Nations Environment Programme 2010).
2. A. Dobson and D. Hayes, 'A Politics of Crisis: Low-Energy Cosmopolitanism', *OpenDemocracy* (2008), http://www.opendemocracy.net/article/a-politics-of-crisis-low-energy-cosmopolitanism.
3. P. Newell and M. Paterson, *Climate Capitalism: Global Warming and the Transformation of the Global Economy* (Cambridge: Cambridge University Press, 2010), 161–6.
4. Fórum Social Mundial/World Social Forum, 'Fórum Social Mundial: Um Outro Mundo É Possível', WSF, http://www.forumsocialmundial.org.br/index.php?cd_language=1.
5. See N. Fenton and N. Couldry, 'Occupy! Rediscovering the General Will in Hard Times', *OpenDemocracy*, 9 January 2012.
6. J. Ballet et al., 'a Note on Sustainability Economics and the Capability Approach', *Ecological Economics* 70, no. 11 (2011); S. Baumgärtner and M. Quaas, 'What Is Sustainability Economics?', *Ecological Economics* 69, no. 3 (2010).
7. J-L. Coraggio, 'Labour Economy', in *The Human Economy: A Citizens Guide*, ed. K. Hart, J-L. Laville, and A.D. Cattani (Cambridge: Polity, 2010), 119.
8. G. Seyfang, 'Shopping for Sustainability: Can Sustainable Consumption Promote Ecological Citizenship', *Environmental Politics* 14, no. 2 (April 2005); 290–306; G. Seyfang, and A. Smith, 'Grassroots Innovation for Sustainable Development: Towards a New Research and Policy Agenda', *Environmental Politics* 16, no. 4 (August 2007): 584–603.
9. D. John, 'Civic Environmentalism', in *Environmental Governance Reconsidered: Challenges, Choices, Opportunities*, ed. R.F. Durant, D.J. Fiorino, and R. O'Leary (Cambridge: MIT Press, 2004), 219; ——, 'Civic Environmentalism', *Issues in Science and Technology* 10, no. 4 (1994), 221.
10. L.C. Hempel, 'Conceptual and Analytical Challenges in Building Sustainable Communities', in *Toward Sustainable Communities: Transition and Transformations in Environmental Policy*, ed. D.A. Mazmanian and M.E. Kraft (Cambridge: MIT Press, 1999), 48.
11. John, 'Civic Environmentalism'; D. John and M. Mlay, 'Community-based environmental protection: Encouraging civic environmentalism', in *Better Environmental Decisions: Strategies for Governments, Businesses and Communities*, ed. K. Sexton et al. (Washington: Island Press, 1999).
12. M. Holden, 'The Tough-Minded and the Tender-Minded: A Pragmatic Turn for Urban Sustainable Development Policy', *Planning Theory and Practice* 9, no. 4 (2008).

13. J. Agyeman and B. Evans, 'Towards Just Sustainability in Urban Communities: Building Equity Rights with Sustainable Solutions', *Annals of the American Academy of Political and Social Sciences* 590 (2003); ——, 'Justice, Governance and Sustainability: Perspectives on Environmental Citizenship from North America and Europe', in *Environmental Citizenship*, ed. A. Dobson and D. Bell (Cambridge: MIT Press, 2006). See also, D.J. Hess, *Localist Movements in a Global Economy: Sustainability, Justice, and Urban Development in the United States* (Cambridge: MIT Press, 2009); R. Sandler and P.C. Pezzullo, eds., *Environmental Justice and Environmentalism: The Social Justice Challenge to the Environmental Movement* (Cambridge: MIT Press, 2007).
14. Agyeman and Evans, 'Justice, Governance and Sustainability: Perspectives on Environmental Citizenship from North America and Europe', 2, 7, 186, 201.
15. J. Agyeman, *Sustainable Communities and the Challenge of Environmental Justice* (New York: New York University Press, 2005), 44.
16. Agyeman and Evans, 'Justice, Governance and Sustainability: Perspectives on Environmental Citizenship from North America and Europe', 190, citing Shutkin.
17. See WPCCC-RME, 'World People's Conference on Climate Change and the Rights of Mother Earth: Building the People's World Movement for Mother Earth', WPCCC-RME, pwccc.wordpress.com.
18. OECD, 'Working together Towards Sustainable Development: The OECD Experience' (Paris: Organization for Economic Co-operation and Development, 2002).
19. D. Schlosberg, *Defining Environmental Justice* (Oxford: Oxford University Press, 2007), 8, 39, 158.
20. P.B. Spahn, 'International Financial Flows and transactions Taxes: Survey and Opinions' (Washington, DC: International Monetary Fund, 1995); J. Tobin, 'A Proposal for International Monetary Reform', *Eastern Economic Journal* 4, no. 3–4 (1978).
21. D. Pimlott, 'Tobin Tax Remains Treasury Ambition', *Financial Times*, 11 December 2009; M. Acharya-Tom Yew, 'Banks Relieved as g20 Backs Off on Bank Tax', *Toronto Star*, 27 June 2010; D. Charter, 'Merkel Leads Calls for Global Financial Tax as Markets Continue to Slide', *The Sunday Times*, 20 May 2010; D. Gow, 'Osborne Wins Wide Support in Fight against Tobin Tax', *The Guardian*, 8 November 2011; N. Wakim, 'Les Altermondialistes d'Attac "Satisfaits" Mais "Pas Dupes" ', *Le Monde*, 16 August 2011.
22. H. Patomakki, *Democratising Globalisation: The Leverage of the Tobin Tax* (London: Zed Books, 2001).
23. B. Cassen, 'On the Attack', *New Left Review* II, no. 19 (2003); ——, 'ATTAC Against the Treaty', *New Left Review* II, no. 33 (2005); S. Spratt, 'A Sterling Solution' (London: Stamp out Poverty/Intelligence Capital Limited, 2006).
24. Newell and Paterson, *Climate Capitalism: Global Warming and the Transformation of the Global Economy*, 10.
25. ATTAC, 'About ATTAC', ATTAC, www.attac.org/en/overview.
26. P.G. Harris, *World Ethics and Climate Change* (Edinburgh: Edinburgh University Press, 2010), 137.
27. L. Elliott, 'Cosmopolitan Environmental Harm Conventions', *Global Society* 20, no. 3 (2006): 357.

28. A. Dobson, *Citizenship and the Environment* (Oxford: Oxford University Press, 2003), 49.
29. S. Dryzek, *Deliberative Global Politics* (Cambridge: Polity, 2006), 4, 136–43.
30. Harris, *World Ethics and Climate Change*, 139, italics in original.
31. P. Edelman, 'A Living Income for Every American: Jobs and Income Strategies for the Twenty-First Century', in *The Social Contract Revisited* (Oxford University: Foundation for Law, Justice and Society, 2007); C. Murray, 'Guaranteed Income as a Replacement for the Welfare State [Foundation for Law, Justice and Society policy brief, Oxford (www.fljs.org)]', *Basic Income Studies* 3, no. 2 (2007); A. Gorz, *Reclaiming Work: Beyond the Wage-Based Society* (Cambridge: Polity, 1999 [1997]); U. Beck, *The Brave New World of Work*, trans. P. Camiller (Cambridge: Polity Press, 2000), 143–5; See http://www.basicincome.org/bien/, Basic Income Studies, available at http://www.bepress.com/bis/.
32. P. Van Parijs, *Arguing for Basic Income* (London: Verso, 1992), 3.
33. E.L. Feige, 'Rethinking Taxation: The Automated Payment Transaction Tax', *Milken Institute Review* First Quarter, no. 11533 (2000); A. Etzioni and A. Platt, 'a Community-Based Guaranteed Income', In *5th Workshop of the Social Contract Revisited* (Oxford: The Foundation for Law, Justice and Society, 2009).
34. S. White, 'Markets, Time and Citizenship', *Renewal* 12, no. 1 (2004); ——, 'Basic Income versus Basic Capital: Can We Resolve the Differences?', *Policy and Politics* 39, no. 167–81 (2011); ——, 'The Republican Critique of Capitalism', *Critical Review of International Social and Political Philosophy* 14, no. 5 (2011): 573–5.
35. Murray, 'Guaranteed Income as a Replacement for the Welfare State [Foundation for Law, Justice and Society Policy Brief, Oxford (www.fljs.org)]'.
36. A. Paz-Fuchs, 'Equality and Personal Responsibility in the New Social Contract', in *Report and Analysis of the 5th Workshop of the Social Contract Revisited* (Oxford University: The Foundation for Law, Justice and Society, 2009), 3; T. Fitzpatrick, 'Basic Income, Post-Productivism and Liberalism', *Basic Income Studies* 4, no. 2, Article 7 (2009); A. Supiot, *Au-Delà De L'emploi: Transformations Du Travail Et Devenir Du Droit Du Travail En Europe. Rapport Pour La Commission Des Communautés Européennes* (Paris: Flammarion, 1999).
37. P-M Boulanger, 'Basic Income and Sustainable Consumption Strategies', *Basic Income Studies* 4, no. 2, Article 5 (2009).
38. J.O. Andersson, 'Basic Income from an Ecological Perspective', *Basic Income Studies* 4, no. 2, Article 4 (2009): 4.
39. Paz-Fuchs, 'Equality and Personal Responsibility in the New Social Contract', 6.
40. This claim has long been made for the guaranteed basic income. See B. Frankel, *The Post-Industrial Utopians* (Cambridge: Polity Press, 1987), 73–86; Boulanger, 'Basic Income and Sustainable Consumption Strategies'; J. Barry, 'Resistance Is Fertile: From Environmental to Sustainability Citizenship', in *Environmental Citizenship*, ed. A. Dobson and D. Bell (Cambridge: MIT Press, 2006); J. Barry, *The Politics of Actually Existing Unsustainability* (Oxford: Oxford University Press, 2012).

41. S. Birnbaum, 'Basic Income, Sustainability and Post-Productivism', *Basic Income Studies* 4, no. 2, Article 3 (2009): 3.
42. T. Princen, *The Logic of Sufficiency* (Cambridge: MIT Press, 2005);——, 'Principles for Sustainability: From Cooperation and Efficiency to Sufficiency', *Global Environmental Politics* 3, no. 1 (2003): 34, 48.
43. ILO, 'Social Protection Floor for a Fair and Inclusive Globalization: Report of the Advisory Group' (Geneva: International Labour Organization with World Health Organization, 2011), 39.
44. EFTA, 'European Fair Trade Association: Joining Fair Trade Forces' (Schin op Geul/Maastricht: European Fair Trade Association, 2006), 1.
45. G. Fridell, 'Fair Trade, Free Trade and the State', *New Political Economy* 15, no. 3 (2010): 467; ——, *Fair Trade Coffee: The Prospects and Pitfalls of Market-Driven Social Justice* (Toronto: Toronto University Press, 2007).
46. ILO, 'ILO Declaration on Social Justice for a Fair Globalization' (Geneva: International Labour Organization, 2008).
47. I.M. Young, 'From Guilt to Solidarity', *Dissent* 50, no. 2 (2003): 40, italics in original.
48. A. Dobson, 'Thick Cosmopolitanism', *Political Studies* 54, no. 1 (2006): 172, 80.
49. M.K. Goodman, D. Maye, and L. Holloway, 'Ethical Foodscapes?: Premises, Promises, Possibilities', *Environment and Planning A* 42, no. 8 (2010); J. Clarke, 'Unsettled Connections: Citizens, Consumers, and the Reform of Public Services', *Journal of Consumer Culture* 7, no. 2 (2007); N. Clarke, 'From Ethical Consumerism to Political Consumption', *Geography Compass* 2, no. 3 (2008); Clarke et al., 'Globalising the consumer: Doing politics in an ethical register'; ——, 'The Political Rationalities of Fair-Trade Consumption in the United Kingdom', *Politics & Society* 35, no. 4 (2007).
50. Goodman, Maye, and Holloway, 'Ethical Foodscapes?', 1788.
51. H. Swarts and I. Bogdan Vasi, 'Which U.S. Cities Adopt Living Wage Ordinances? Predictors of Adoption of a New Labor Tactic', *Urban Affairs Review* 47, no. 6 (2011).
52. C. Barnett, P. Cloke, N. Clarke, and A. Malpass, 'Consuming Ethics: Articulating the Subjects and Spaces of Ethical Consumption', *Antipode* 37, no. 1 (2005): 41, italics added.
53. See I.M. Young, *Responsibility for Justice* (Oxford: Oxford University Press, 2011), 104–13.
54. A. Dobson, 'Citizens, Citizenship and Governance for Sustainability', in *Governing Sustainability*, ed. W.N. Adger and A. Jordan (Cambridge: Cambridge University Press, 2009), 40, 136; J. Barry, 'Resistance Is Fertile: From Environmental to Sustainability Citizenship', in *Environmental Citizenship*, ed. A. Dobson and D. Bell (Cambridge: MIT Press, 2006), 28.
55. C. Barnett, 'Ways of Relating: Hospitality and the Acknowledgement of Otherness', *Progress in Human Geography* 29, no. 1 (2005). cited in, Barnett et al., 'Consuming Ethics: Articulating the Subjects and Spaces of Ethical Consumption', 42.
56. Dobson, 'Thick Cosmopolitanism', 172–3.
57. H.E. Daly, *Ecological Economics and Sustainable Development: Selected Essays of Herman Daly*, ed. J.C.J.M. Van den Bergh (Northampton: Edward Elgar, 2007), 52–3.

58. United Nations, 'Agenda 21 Earth Summit: United Nations program of action from Rio' (New York: United Nations 1992).
59. B. Moldan and A. Lyon Dahl, 'Challenges to Sustainability Indicators', in *SCOPE 67: Sustainability Indicators A Scientific Assessment*, ed. T. Hak, B. Moldan, and A. Lyon Dahl (Washington, DC: Island Press, 2007).
60. UN, 'General Assembly Resolution 2 Session 55 United Nations Millennium Declaration' (New York: United Nations, 2000); OECD, 'Better Life Index', Organization for Economic Co-operation and Development, http://www.oecdbetterlifeindex.org/.
61. Y. Rydin, N. Holman, and E. Wolff, 'Local Sustainability Indicators', *Local Environment* 8, no. 6 (2003): 581.
62. T. Rutland and A. Aylett, 'The Work of Policy: Actor Networks, Governmentality, and Local Action on Climate Change in Portland, Oregon', *Environment and Planning D: Society and Space* 26 (2008): 621–3.
63. K. Eckerberg and E. Mineur, 'The Use of Local Sustainability Indicators: Case Studies in Two Swedish Municipalities', *Local Environment* 8, no. 6 (2003); M. Holden, 'Community Interests and Indicator System Success', *Social Indicators Research* 92 (2009); P. McKinlay, 'The Challenge of Democratic Participation in the Community Development Process', *Community Development Journal* 41, no. 4 (2006); R.M. Pollock and G.S. Whitelaw, 'Community-Based Monitoring in Support of Local Sustainability', *Local Environment* 10, no. 3 (2005).
64. C.M. Hendriks, 'Deliberative Governance in the Context of Power', *Policy and Society* 28 (2009): 174.
65. Holden, 'The Tough-Minded and the Tender-Minded: A Pragmatic Turn for Urban Sustainable Development Policy'.
66. L. Boltanski and E. Chiapello, *The New Spirit of Capitalism*, trans. G. Elliott (London: Verso Books, 2005 [1999]), 107, 139.
67. N. Holman, 'Incorporating Local Sustainability Indicators into Structures of Local Governance: A Review of the Literature', *Local Environment* 14, no. 4 (2009): 372; M. Holden, 'Urban Indicators and the Integrative Ideals of Cities', *Cities* 23, no. 3 (2006); ——, 'Revisiting the Local Impact of Community Indicators Projects: Sustainable Seattle as a Prophet in Its Own Land', *Applied Research in Quality of Life* 1 (2006).
68. C.M. Hendriks, J.S. Dryzek, and C. Hunold, 'Turning Up the Heat: Partisanship in Deliberative Innovation', *Political Studies* 55 (2007): 378.
69. Victorian Local Government Association, 'Liveable & Just Toolkit: Connecting Communities, Strengthening Democracy', VLGA and the McCaughey Centre at the University of Melbourne, http://www.vlga.org.au/Resources/Liveable_Just_Toolkit.aspx; Vancouver Foundation, 'Vital Signs for Metro Vancouver', Vancouver Foundation and Metro Vancouver Regional Authority, http://www.vancouverfoundationvitalsigns.ca/.
70. J. Becker, 'Measuring Progress Towards Sustainable Development: An Ecological Framework for Selecting Indicators', *Local Environment* 10, no. 1 (2005); P. McAlpine and A. Birnie, 'Is There a Correct Way of Establishing Sustainability Indicators? The Case of Sustainability Indicator Development on the Island of Guernsey', *Local Environment* 10, no. 3 (2005); Pollock and Whitelaw, 'Community-Based Monitoring in Support of Local Sustainability'.

71. C. Miller, 'New Civic Epistemologies of Quantification: Making Sense of Indicators of Local and Global Sustainability', *Science, Technology and Human Values* 30, no. 3 (2005): 430–1.
72. Sustainable Seattle, 'Sustainable Seattle and Indicators', King County, http://www.sustainableseattle.org/Programs/RegionalIndicators/index_html; ——, 'Indicators Principles', Sustainable Seattle, www.sustainableseattle.org; D. May, 'Uses of Community Accounts' (St Johns: Newfoundland and Labrador Statistics Agency/ Memorial University of Newfoundland, 2005); S. Cantin, R. Rogers, and S. Burdett, 'Indicators of Healthy and Vibrant Communities Roundtable' (Toronto: Canadian Policy Research Networks/ The Ontario Trillium Foundation, 2008); M. Cooper, 'Social Sustainability in Vancouver', *Canadian Policy Research Network* F, no. 62 (2006); M. Holden and C. Mochrie, 'The Regional Vancouver Urban Observatory (RVU): Counting on Vancouver, "Our View" of the Region', *Revista Internacional de Sostenabilidad, Tecnologia y Humanismo* 1 (2006); ——, 'Counting on Vancouver: Our View of the Region. Inaugural Report of the Regional Vancouver Urban Observatory' (Vancouver: Regional Vancouver Urban Observatory/Simon Fraser University Urban Studies Program 2006); J. Wiseman et al., 'Measuring Wellbeing, Engaging Communities', VicHealth, Victoria University, http://www.communityindicators.net.au; J. Humm, K. Jones, and G. Chanan, 'Testing Indicators of Community Involvement: Final Report' (London: Community Development foundation, 2005).
73. D. Torgerson, 'Democracy through Policy Discourse', in *Deliberative Policy Analysis: Understanding Governance in the Network Society*, ed. M. Hajer and H. Wagenaar (Cambridge: Cambridge University Press, 2003), 128.
74. Dryzek, *Deliberative Global Politics*: 47.
75. 'A Frame in the Fields: Policymaking and the Reinvention of Politics', in *Deliberative Policy Analysis: Understanding Governance in the Network Society*, ed. M. Hajer and H. Wagenaar (Cambridge: Cambridge University Press, 2003), 95–6, italics in original.

7 Conclusion

1. A. Dobson, *Green Political Thought*, 4th ed. (London: Routledge, 2007), 42.
2. Ibid., 190.
3. R.E. Goodin, *Green Political Theory* (Cambridge: Polity, 1992), 24–5.
4. L. Dumont, *Essays on Individualism: Modern Ideology in Anthropological Perspective*, English ed. (Chicago: University of Chicago Press, 1986 [1983]), 217.
5. D. Schlosberg, *Defining Environmental Justice* (Oxford: Oxford University Press, 2007), 8.
6. N. Fraser, *Scales of Justice: Reimagining Political Space in a Globalizing World* (New York: Columbia University Press, 2009), 15.
7. Dobson, 'Thick Cosmopolitanism'; J. Barry, *The Politics of Actually Existing Unsustainability* (Oxford: Oxford University Press, 2012).
8. P. Rosanvallon, *Democratic Legitimacy: Impartiality, Reflexivity, Proximity*, trans. A. Goldhammer (Princeton: Princeton University Press, 2011 [2008]), 9.
9. Fraser, *Scales of Justice: Reimagining Political Space in a Globalizing World*, 57.

Bibliography

Abercrombie, N., S. Hill, and B.S. Turner. *Sovereign Individuals of Capitalism*. London: Allen & Unwin, 1986.

Acharya-Tom Yew, M. 'Banks Relieved as G20 Backs Off on Bank Tax.' *Toronto Star*, 27 June 2010.

Adams, M., and J. Raisborough. 'What Can Sociology Say About Fair Trade? Class, Reflexivity and Ethical Consumption.' *Sociology* 42, no. 6 (2008): 1165–82.

Adorno, T., and M. Horkheimer. *Dialectic of Enlightenment*. London: Allen Lane Publishers, 1973 [1947].

Agyeman, J. *Sustainable Communities and the Challenge of Environmental Justice*. New York: New York University Press, 2005.

Agyeman, J., and B. Evans. 'Justice, Governance and Sustainability: Perspectives on Environmental Citizenship from North America and Europe.' In *Environmental Citizenship*, edited by A. Dobson and D. Bell. 185–206. Cambridge: MIT Press, 2006.

——. *The Performance of Politics: Obama's Victory and the Democratic Struggle for Power*. Oxford: Oxford University Press, 2010.

——. 'Towards Just Sustainability in Urban Communities: Building Equity Rights with Sustainable Solutions.' *Annals of the American Academy of Political and Social Sciences* 590 (2003): 35–53.

Alston, P., ed. *Labour Rights as Human Rights*. Oxford: Oxford University Press, 2005.

Andersson, J.O. 'Basic Income from an Ecological Perspective.' *Basic Income Studies* 4, no. 2, Article 4 (2009): 1–8.

Arendt, H. *The Human Condition*. 2nd ed. Chicago: University of Chicago Press, 1958.

Arias-Maldonado, M. 'The Democratisation of Sustainability: The Search for a Green Democratic Model.' *Environmental Politics* 9, no. 4 (2000): 43–58.

Atkinson, G. 'Measuring Corporate Sustainability.' *Journal of Environmental Planning and Management* 43, no. 2 (2000): 235–52.

ATTAC. 'About ATTAC.' ATTAC, http://www.attac.org/en/overview.

Bachrach, P. *The Theory of Democratic Elitism*. London: University of London Press, Ltd., 1967.

Ballet, J., D. Bazin, J-L. Dubois, and F-R. Mahieu. 'A Note on Sustainability Economics and the Capability Approach.' *Ecological Economics* 70, no. 11 (2011): 1831–4.

Banerjee, S.B. 'Corporate Social Responsibility: The Good, the Bad and the Ugly.' *Critical Sociology* 34, no. 1 (2008): 51–79.

Barnett, C. 'Ways of Relating: Hospitality and the Acknowledgement of Otherness.' *Progress in Human Geography* 29, no. 1 (2005): 5–21.

Barnett, C., P. Cloke, N. Clarke, and A. Malpass. 'Consuming Ethics: Articulating the Subjects and Spaces of Ethical Consumption.' *Antipode* 37, no. 1 (2005): 23–45.

Barry, J. 'Ecological Modernization.' In *Debating the Earth*, edited by J.S. Dryzek and D. Schlosberg. 303–22. Oxford: Oxford University Press, 2005.
——. *The Politics of Actually Existing Unsustainability*. Oxford: Oxford University Press, 2012.
——. 'Resistance Is Fertile: From Environmental to Sustainability Citizenship.' In *Environmental Citizenship*, edited by A. Dobson and D. Bell. 21–48. Cambridge: MIT Press, 2006.
——. *Rethinking Green Politics*. London: Sage, 1999.
——. 'Review: Denaturalizing Ecological Politics: Alienation from Nature from Rousseau to the Frankfurt School and Beyond by Andrew Biro.' *Environmental Politics* 17, no. 4 (2007): 688–9.
——. 'Sustainability, Political Judgement and Citizenship: Connecting Green Politics and Democracy.' In *Democracy and Green Political Thought*, edited by B. Doherty and M. de Geus. 115–31. London: Routledge, 1996.
Barry, J., and P. Doran. 'Refining Green Political Economy: From Ecological Modernisation to Economic Security and Sufficiency.' *Analyse & Kritik* 28, no. 2 (2006): 250–75.
Barry, J., and R. Eckersley. 'W(H)Ither the Green State?.' In *The State and the Global Ecological Crisis*, edited by J. Barry and R. Eckersley. 255–72. Cambridge: MIT Press, 2005.
Barry, J., and K. Smith. 'Civic Republicanism and Green Politics.' Chap. 13. In *Building a Citizen Society*, edited by S. White and D. Leighton. 137–46. London: Lawrence & Wishart, 2008.
Baudrillard, J. *For a Critique of the Political Economy of the Sign*. Translated by C. Levin. St. Louis: Telos Press, 1981 [1972].
Bauman, Z. *Consuming Life*. London: Polity Press, 2007.
——. *The Individualized Society*. Cambridge: Polity Press, 2001.
——. 'The Poor – and the Rest of Us.' *Arena* New Series, no. 12 (1998): 43–66.
——. 'Social Issues of Law and Order.' *British Journal of Criminology* 40 (2000): 205–21.
——. *Work, Consumerism and the New Poor*. Buckingham: Open University Press, 1998.
Baumgärtner, S., and M. Quaas. 'What Is Sustainability Economics?.' *Ecological Economics* 69, no. 3 (2010): 445–50.
Beck, U. *The Brave New World of Work*. Translated by P. Camiller. Cambridge: Polity Press, 2000.
——. *Ecological Politics in an Age of Risk*. Cambridge: Polity Press, 1995.
——. 'Freedom's Children.' Translated by P. Camiller. Chap. 12 In *Individualization*, edited by U. Beck and E. Beck-Gernsheim. 156–72. London: Sage, 2001.
——. 'On the Way to the Industrial Risk-Society? Outline of an Argument.' *Thesis Eleven* 23 (1989): 86–103.
——. *The Reinvention of Politics: Rethinking Modernity in the Global Social Order*. Cambridge: Polity Press, 1997.
——. 'The Reinvention of Politics: Towards a Theory of Reflexive Modernisation.' In *Reflexive Modernity*, edited by A. Giddens, S. Lash, and U. Beck. 1–55. London: Polity Press, 1997.
——. *Risk Society: Towards a New Modernity*. London: Sage Publishers, 1992.
——. 'The Silence of Words and Political Dynamics in the World Risk Society.' *Logos* 1, no. 4 (2002): 1–18.

——. *World Risk Society*. London: Polity Press/Blackwell Books, 1999.

Beck, U., and E. Beck-Gernsheim. *Individualization*. Translated by P. Camiller. Theory, Culture & Society. edited by M. Featherstone. London: Sage Publications, 2002.

Beck, U., and J. Rutherford. 'Zombie Categories: An Interview with Ulrich Beck.' In *Individualization*, edited by U. Beck and E. Beck-Gernsheim. 202–13. London: Sage, 2002.

Becker, J. 'Measuring Progress Towards Sustainable Development: An Ecological Framework for Selecting Indicators.' *Local Environment* 10, no. 1 (2005): 87–101.

Behr, R. 'Anti-Consumerism: New Frontier or Dead End for Progressive Politics?.' *Public Policy Research* 17, no. 2 (2010): 123–9.

Bell, D. *The Coming of Postindustrial Society*. New York: Basic Books, 1973.

——. *The Cultural Contradictions of Capitalism*. 2nd ed. London: Heinemann Books, 1978.

——. *The End of Ideology*. New York: Glencoe Free Press, 1960.

Bell, D.R. 'Liberal Environmental Citizenship.' *Environmental Politics* 14, no. 2 (2005): 179–94.

Benhabib, S., ed. *Democracy and Difference: Contesting Boundaries of the Political*. Princeton: Princeton University Press, 1996.

Benioff, M., and K. Southwick. *Compassionate Capitalism: How Corporations Can Make Doing Good an Integral Part of Doing Well*. New York: Career Press, 2003.

Berman, M. *All That Is Solid Melts into Air: The Experience of Modernity*. New York: Simon & Schuster, 1982.

Birnbaum, S. 'Introduction: Basic Income, Sustainability and Post-Productivism.' *Basic Income Studies* 4, no. 2, Article 3 (2009): 1–7.

Biro, A. *Denaturalizing Ecological Politics: Alienation from Nature from Rousseau to the Frankfurt School and Beyond*. Toronto: University of Toronto Press, 2005.

Blühdorn, I. 'Beyond Criticism and Crisis: On the Post-Critical Challenge of Niklas Luhmann.' *Debatte* 7, no. 2 (1999): 185–99.

——. 'Democracy, Efficiency, Futurity: Contested Objectives of Societal Reform.' In *Economic Efficiency – Democratic Empowerment: Contested Modernization in Britain and Germany*, edited by I. Blühdorn and I. Jun. 69–98. Lanham: Rowman & Littlefield, 2007.

——. 'An Offer One Might Prefer to Refuse: The Systems Theoretical Legacy of Niklas Luhmann.' *European Journal of Social Theory* 3, no. 3 (2000): 339–54.

——. *Post-Ecologist Politics: Social Theory and the Abdication of the Ecologist Paradigm*. London: Routledge, 2000.

——. 'Self-Experience in the Theme Park of Radical Action? Social Movement and Political Articulation in the Late-Modern Condition.' *European Journal of Social Theory* 9, no. 1 (2006): 23–42.

——. 'Sustaining the Unsustainable: Symbolic Politics and the Politics of Simulation.' *Environmental Politics* 16, no. 2 (2007): 251–75.

Blühdorn, I., and I. Welsh. 'Eco-Politics Beyond the Paradigm of Sustainability: A Conceptual Framework and Research Agenda.' *Environmental Politics* 16, no. 2 (2007): 185–205.

Bobbio, N. *Left and Right: The Significance of a Political Distinction*. Translated by A. Cameron. Cambridge: Polity, 1996 [1995].

Boggs, C. *The End of Politics: Corporate Power and the Decline of the Public Sphere*. New York: Guilford Press, 2000.
Boles, N. *Which Way's Up: The Future for Coalition Britain and How to Get There*. London: Biteback Publishers, 2010.
Boltanski, L., and E. Chiapello. *The New Spirit of Capitalism*. Translated by G. Elliott. London: Verso Books, 2005 [1999].
Boltanski, L., and L. Thévenot. *On Justification: Economies of Worth*. Translated by C. Porter. Princeton Studies in Cultural Sociology. edited by P. DiMaggio, M. Lamont, R.J. Wuthnow, and V. Zelizer. Princeton: Princeton University Press, 2006 [1991].
——. 'The Sociology of Critical Capacity.' *European Journal of Social Theory* 2, no. 3 (1999): 359–77.
Boulanger, P-M. 'Basic Income and Sustainable Consumption Strategies.' *Basic Income Studies* 4, no. 2, Article 5 (2009): 1–11.
Brenner, N. *Spaces of Neoliberalism: Urban Restructuring in North America and Western Europe*. London: Wiley-Blackwell, 2003.
Brenner, R. *The Boom and the Bubble*. 2nd ed. London: Verso, 2005.
——. 'New Boom or New Bubble?.' *New Left Review* II, no. 25 (January–February 2004): 57–102.
Brown, H.S., M. de Jong, and T. Lessidrenska. 'The Rise of the Global Reporting Initiative: A Case of Institutional Entrepreneurship.' *Environmental Politics* 18, no. 2 (2009): 182–200.
Brown, K. 'Human Development and Environmental Governance: A Reality Check.' In *Governing Sustainability*, edited by W.N. Adger and A. Jordan. 32–51. Cambridge: Cambridge University Press, 2009.
Buckler, S., and D.P. Dolowitz. 'Theorizing the Third Way: New Labour and Social Justice.' *Journal of Political Ideologies* 5, no. 3 (2000): 301–20.
Bulkeley, H., and A.P.J. Mol. 'Participation and Environmental Governance: Consensus, Ambivalence and Debate.' *Environmental Values* 12, no. 2 (2003): 143–54.
Bunting, M. *Willing Slaves: How the Overwork Culture Is Ruling Our Lives*. London: HarperCollins Publishers, 2004.
Burgess, J., C.M. Harrison, and P. Filius. 'Environmental Communication and the Cultural Politics of Citizenship.' *Environment and Planning A* 30, no. 8 (1998): 1445–60.
Burkitt, B., and F. Ashton. 'The Birth of the Stakeholder Society.' *Critical Social Policy* 16, no. 49 (1996): 3–16.
Callinicos, A. *Equality*. Themes for the 21st Century. Cambridge: Polity Press, 2000.
Campbell, C. *The Romantic Ethic and the Spirit of Consumerism*. Oxford: Blackwell, 1987.
Campbell, D. 'Social Messages Now in Fashion.' *The Guardian Weekly*, 3–9 October 2002, 22.
Cantin, S., R. Rogers, and S. Burdett. 'Indicators of Healthy and Vibrant Communities Roundtable.' 29. Toronto: Canadian Policy Research Networks/ The Ontario Trillium Foundation, 2008.
Casey, N. 'Melbourne Benchmarking and Liveability.' Melbourne: City of Melbourne, 2009.

——. 'Melbourne City Research: International City Comparisons.' Melbourne: City of Melbourne, 2011.

Cassen, B. 'ATTAC against the Treaty.' *New Left Review* II, no. 33 (2005): 27–33.

——. 'On the Attack.' *New Left Review* II, no. 19 (2003): 41–60.

Castells, M. 'Materials for an Exploratory Theory of the Network Society.' *British Journal of Sociology* 51, no. 1 (2000): 5–24.

Castoriadis, C. *The Imaginary Institution of Society*. Translated by K. Blamey. Cambridge: Polity Press, 1987 [1975].

Charter, D. 'Merkel Leads Calls for Global Financial Tax as Markets Continue to Slide.' *The Sunday Times*, 20 May 2010.

Chiapello, E. 'Reconciling the Two Principal Meanings of the Notion of Ideology: The Example of the Concept of the 'New Spirit of Capitalism.' *European Journal of Social Theory* 6, no. 2 (2003): 155–71.

Christoff, P. 'Ecological Citizens and Ecologically Guided Democracy.' In *Democracy and Green Political Thought: Sustainability, Rights and Citizenship*, edited by B. Doherty and M. de Geus. 151–69. London: Routledge, 1996.

——. 'Ecological Modernisation, Ecological Modernities.' *Environmental Politics* 5, no. 3 (1996): 476–500.

——. 'Out of Chaos, a Shining Star? Toward a Typology of Green States.' In *The State and the Global Ecological Crisis*, edited by J. Barry and R. Eckersley. 75–96. Cambridge: MIT Press, 2005.

Clarke, J. 'New Labour's Citizens: Activated, Empowered, Responsibilized, Abandoned?.' *Critical Social Policy* 25, no. 4 (2005): 447–63.

——. 'Unsettled Connections: Citizens, Consumers, and the Reform of Public Services.' *Journal of Consumer Culture* 7, no. 2 (2007): 159–78.

Clarke, N. 'From Ethical Consumerism to Political Consumption.' *Geography Compass* 2, no. 3 (2008): 1870–84.

Clarke, N., C. Barnett, P. Cloke, and A. Malpass. 'Globalising the Consumer: Doing Politics in an Ethical Register.' *Political Geography* 26, no. 3 (2007): 231–49.

——. 'The Political Rationalities of Fair-Trade Consumption in the United Kingdom.' *Politics & Society* 35, no. 4 (Dec. 2007): 583–607.

Cooper, M. 'Social Sustainability in Vancouver.' *Canadian Policy Research Network* F, no. 62 (2006).

Coraggio, J-L. 'Labour Economy.' In *The Human Economy: A Citizens Guide*, edited by K. Hart, J-L. Laville, and A.D. Cattani. 119–29. Cambridge: Polity, 2010.

Crozier, M., S.P. Huntington, and J. Watanuki. *The Crisis of Democracy: Report on the Governability of Democracies to the Trilateral Commission*. New York: NYU Press, 1975.

Daly, H. *Ecological Economics and the Ecology of Economics*. Cheltenham: Edward Elgar, 1999.

Daly, H.E. *Ecological Economics and Sustainable Development: Selected Essays of Herman Daly*. Advances in Ecological Economics. edited by J.C.J.M. Van den Bergh Northampton: Edward Elgar, 2007.

Daly, H.E., and J.B. Cobb Jr. *For the Common Good: Redirecting the Economy Towards Community, the Environment and Sustainable Development*. London: GreenPrint, 1990.

Davis, M. 'The Political Economy of Late Imperial America.' *New Left Review* 143 (1984): 8–36.

Deakin, S., and F. Wilkinson. *The Law of the Market: Industrialization, Employment, and Legal Evolution*. Oxford: Oxford University Press, 2005.

Dean, H. 'Popular Discourse and the Ethical Deficiency of 'Third Way' Conceptions of Citizenship.' *Citizenship Studies* 8, no. 1 (2004): 65–82.

Debray, R. 'A Modest Contribution to the Rites and Ceremonies of the Tenth Anniversary.' *New Left Review* I, no. 115 (May–June 1979): 44–65.

Delanty, G. *Citizenship in a Global Age: Society, Culture, Politics*. Buckingham: Open University Press, 2000.

——. *Modernity and Postmodernity: Knowledge, Power, and the Self*. London: Sage Publications, 2000.

Desrosières, A. 'Historiciser L'action Publique: L'etat, Le Marche Et Les Statistiques.' In *Historicites De L'action Publique*, edited by P. Laborier and D. Trom. 207–21. Paris: Presses Universitaires de France, 2003.

Diamond, J. *Guns, Germs and Steel: A Short History of Everybody for the Last 13 000 Years*. London: Vintage Books, 1998.

Dicken, P. *Global Shift: Transforming the World Economy*. 3rd ed. London: Sage Books, 1999. Paul Chapman Publishing Ltd., London.

Dobson, A. 'Book Review: "The Politics of Climate Change".' *Environmental Politics* 19, no. 2 (2010): 310–12.

——. 'Citizens, Citizenship and Governance for Sustainability.' In *Governing Sustainability*, edited by W.N. Adger and A. Jordan. 125–41. Cambridge: Cambridge University Press, 2009.

——. *Citizenship and the Environment*. Oxford: Oxford University Press, 2003.

——. 'Citizenship.' In *Political Theory and the Ecological Challenge*, edited by A. Dobson and R. Eckersley. 216–31. Cambridge: Cambridge University Press, 2006.

——. 'Ecocentrism: A Response to Paul Kingsnorth.' In, *OpenDemocracy* (2010). Published electronically 17 August, http://www.opendemocracy.net/andrew-dobson/ecocentrism-response-to-paul-kingsnorth.

——. 'Ecological Citizenship: A Disruptive Influence?.' In *Politics at the Edge: The PSA Yearbook 1999*, edited by C. Pierson and S. Tormey. 40–61. New York: Political Studies Association/Macmillan, 1999.

——. *Green Political Thought*. 4th ed. London: Routledge, 2007.

——. 'Thick Cosmopolitanism.' *Political Studies* 54, no. 1 (2006): 165–84.

Dobson, A., and D. Bell, eds. *Environmental Citizenship*. Cambridge: MIT Press, 2006.

Dobson, A., and D. Hayes. 'A Politics of Crisis: Low-Energy Cosmopolitanism.' In, *OpenDemocracy* (2008). Published electronically 26 October, http://www.opendemocracy.net/article/a-politics-of-crisis-low-energy-cosmopolitanism.

Dobson, A., and Á. Valencia Sáiz. 'Introduction.' In *Citizenship, Environment, Economy*. 157–62. London: Routledge, 2005.

Drache, D., and M. Getler, eds. *The New Era of Global Competition: State Policy and Market Power*. Montreal: McGill-Queen's University Press, 1991.

Dryzek, J.S. *Deliberative Democracy and Beyond: Liberals, Critics, Contestations*. Oxford: Oxford University Press, 2000.

——. *Deliberative Global Politics*. Cambridge: Polity, 2006.

——. *Politics of the Earth*. 2nd ed. Oxford: Oxford University Press, 2004.

Dryzek, J.S., and P. Dunleavy. *Theories of the Democratic State*. New York: Palgrave Macmillan, 2009.

Dryzek, J.S., D. Downs, H-K. Hernes, and D. Schlosberg. *Green States and Social Movements: Environmentalism in the United States, United Kingdom, Germany and Norway*. Oxford: Oxford University Press, 2003.
du Gay, P. *Consumption and Identity at Work*. London: Sage Books, 1996.
Duménil, G., and D. Lévy. *Capital Resurgent: Roots of the Neoliberal Revolution*. Translated by D. Jeffers. Cambridge: Harvard University Press, 2004.
——. 'Neoliberal Income Trends.' *New Left Review* II, no. 30 (November–December 2004): 105–33.
Dumont, L. *Essays on Individualism: Modern Ideology in Anthropological Perspective*. English ed. Chicago: University of Chicago Press, 1986 [1983].
——. *From Mandeville to Marx: The Genesis and Triumph of Economic Ideology*. Chicago: University of Chicago Press, 1977.
——. *German Ideology: From France to Germany and Back*. Chicago: Chicago University Press, 1994 [1991].
——. 'Sur L'ideologie Politique Française: Une Perspective Comparative.' *Le Debat* 58 (1980): 116–42.
Dumont, L., and C. Delacampagne. 'Louis Dumont and the Indian Mirror.' *RAIN* 43 (1981): 4–7.
Duncan, G. 'After Happiness.' *Journal of Political Ideologies* 12, no. 1 (2007): 85–108.
Dunlap, R., and G.A. Mertig. 'Global Concern for the Environment: Is Affluence a Prerequisite?.' *Journal of Social Issues* 51, no. 3 (1995): 121–37.
——. 'Global Environmental Concern: An Anomaly for Postmaterialism.' *Social Science Quarterly* 78, no. 1 (1997): 24–9.
Dwyer, P. 'Conditional Citizens? Welfare Rights and Responsibilities in the Late 1990s.' *Critical Social Policy* 18, no. 4 (1998): 493–517.
Eagleton, T. *The Illusions of Postmodernism*. London: Blackwell Publishing, 1996.
Eckerberg, K., and E. Mineur. 'The Use of Local Sustainability Indicators: Case Studies in Two Swedish Municipalities.' *Local Environment* 8, no. 6 (2003): 591–614.
Eckersley, R. *The Green State: Rethinking Democracy and Sovereignty*. Cambridge: MIT Press, 2004.
Economist Intelligence Unit. 'Liveability Ranking.' The Economist, http://www.economist.com/markets/rankings/displaystory.
Edelman, P. 'A Living Income for Every American: Jobs and Income Strategies for the Twenty-First Century.' In *The Social Contract Revisited*. Oxford University: Foundation for Law, Justice and Society, 2007.
Eder, K. *The New Politics of Class: Social Movements and Cultural Dynamics in Advanced Societies*. London: Sage, 1993.
EFTA. 'European Fair Trade Association: Joining Fair Trade Forces.' Schin op Geul/Maastricht: European Fair Trade Association, 2006.
Eggers, B. *Government by Network*. Washington: Brookings Institute Press, 2005.
Ehrenreich, B. *Bait and Switch: The Futile Pursuit of the Corporate Dream*. London: Granta, 2006.
——. *Nickeld and Dimed: On (Not) Getting by in America*. New York: Metropolitan Press, 2002.
Elkington, J. *Cannibals with Forks: The Triple Bottom-Line of 21st Century Business* Gabriola Island: New Society Publishers, 1997.
——. *The Chrysalis Economy*. Oxford: Capstone Press, 2001.

Elliott, A., and C. Lemert. *The New Individualism: The Emotional Costs of Globalization*. London: Routledge, 2006.
Elliott, L. 'Cosmopolitan Environmental Harm Conventions.' *Global Society* 20, no. 3 (2006): 345–63.
——. 'Global Environmental Governance.' In *Global Governance: Critical Perspectives*, edited by R. Wilkinson and S. Hughes. 57–74. London: Routledge, 2002.
Espeland, W., and M. Sauder. 'Rankings and Reactivity: How Public Measures Recreate Social Worlds.' *American Journal of Sociology* 113, no. 1 (2007): 1–40.
Esping-Andersen, G. *The Three Worlds of Welfare Capitalism*. Princeton: Princeton University Press, 1990.
Etzioni, A., and A. Platt. 'A Community-Based Guaranteed Income.' In *5th Workshop of the Social Contract Revisited*. Oxford: The Foundation for Law, Justice and Society, 2009.
Featherstone, M. 'Automobilities: Introduction to the Special Edition.' *Theory, Culture & Society* 21, no. 4/5 (2004): 1–24.
——. *Consumer Culture and Postmodernism*. Theory, Culture & Society. edited by M. Featherstone. 2nd ed. London: Sage Publishers, 1996.
Feige, E.L. 'Rethinking Taxation: The Automated Payment Transaction Tax.' *Milken Institute Review* First Quarter, no. 11533 (2000): 42–53.
Fenton, N., and N. Couldry. 'Occupy: Rediscovering the General Will in Hard Times.' *OpenDemocracy*, 9 January 2012.
Fischer, F. *Democracy & Expertise: Reorienting Policy Inquiry*. Oxford: Oxford University Press, 2009.
——. *Reframing Public Policy: Discursive Politics and Deliberative Processes*. Oxford: Oxford University Press, 2003.
Fitzpatrick, T. 'Basic Income, Post-Productivism and Liberalism.' *Basic Income Studies* 4, no. 2, Article 7 (2009): 1–11.
Florida, R. *The Rise of the Creative Class*. New York: Basic Books, 2002.
Fórum Social Mundial/World Social Forum. 'Fórum Social Mundial: Um Outro Mundo É Possível.' WSF, http://www.forumsocialmundial.org.br/index.php?cd_language=1.
Frank, T. *The Conquest of Cool: Business Culture, Counterculture, and the Rise of Hip Consumerism*. Chicago: University of Chicago Press, 1997.
——. *What's the Matter with Kansas?* New York: Vintage Books, 2004.
Frankel, B. *The Post-Industrial Utopians*. Cambridge: Polity Press, 1987.
Fraser, N. 'Feminism, Capitalism and the Cunning of History.' *New Left Review* II, no. 56 (2009): 97–117.
——. *Justice Interruptus: Critical Reflections on the Postsocialist Condition*. London: Routledge, 1997.
——. *Scales of Justice: Reimagining Political Space in a Globalizing World*. New York: Columbia University Press, 2009.
——. *Unruly Practices: Power, Discourse and Gender in Contemporary Social Theory*. Cambridge: Polity Press, 1989.
Fraser, N., and L. Gordon. 'Contract Versus Charity: Why Is There No Social Citizenship in the United States.' *Socialist Review* 23, no. 3 (1992): 45–68.
Fraser, N., and A. Honneth. *Redistribution or Recognition? A Political-Philosophical Exchange*. Translated by J. Golb, J. Ingram, and C. Wilke. London: Verso Books, 2003.

Freedland, M. *The Personal Employment Contract*. Oxford: Oxford University Press, 2003.
Fridell, G. *Fair Trade Coffee: The Prospects and Pitfalls of Market-Driven Social Justice*. Toronto: Toronto University Press, 2007.
——. 'Fair Trade, Free Trade and the State.' *New Political Economy* 15, no. 3 (2010): 457–70.
Fritsch, S. 'The UN Global Compact and the Global Governance of Corporate Social Responsibility: Complex Multilateralism for a More Human Globalisation?.' *Global Society* 22, no. 1 (2008): 1–26.
Gabrielson, T. 'Green Citizenship: A Review and Critique.' *Citizenship Studies* 12, no. 4 (2008): 429–46.
Gabrielson, T., and K. Parady. 'Corporeal Citizenship: Rethinking Green Citizenship through the Body.' *Environmental Politics* 19, no. 3 (2010): 374–91.
Galbraith, J.K. *The Affluent Society*. London: Hamish Hamilton, 1958.
Gauchet, M. 'Democratic Pacification and Civic Desertion.' *Thesis Eleven* 29 (1991): 1–13.
——. *The Disenchantment of the World: A Political History of Religion*. Translated by O. Burge. New French Thought. edited by T. Pavel and M. Lilla. Princeton: Princeton University Press, 1997 [1985].
——. 'A New Age of Personality: An Essay on the Psychology of Our Times.' *Thesis Eleven* 60 (2000): 23–41.
Gauchet, M., B. Renouvin, and S. Rothnie. 'Democracy and the Human Sciences: An Interview with Marcel Gauchet.' *Thesis Eleven* 26 (1990 [1988]): 140–50.
General Electric Company. 'Ongoing Stakeholder Engagement.' GE, http://www.ge.com/citizenship/programs-activities/business-processes/ongoing-stakeholder-engagement.html.
Gereffi, G., R. Garcia-Johnson, and E. Sasser. 'The NGO-Industrial Complex.' *Foreign Policy* 125 (July–August 2001): 56–65.
Geuss, R. *Outside Ethics*. Princeton: Princeton University Press, 2005.
——. *Philosophy and Real Politics*. Princeton: Princeton University Press, 2008.
Giddens, A. *The Class Structure of the Advanced Societies*. Sociology. edited by P.S. Cohen London: Hutchinson & Co. Ltd, 1974.
——. *The Consequences of Modernity*. London: Polity Press, 1990.
——. *Modernity & Self Identity*. Cambridge: Polity Press, 1991.
——. *The Politics of Climate Change*. Cambridge: Polity Press, 2009.
——. 'The Politics of Climate Change: National Responses to the Challenge of Global Warming.' 20. London: Policy Network, 2008.
——. *The Third Way*. Cambridge: Polity Press, 1998.
——. *The Third Way and Its Critics*. Cambridge: Cambridge University Press, 2000.
Giddens, A., S. Lash, and U. Beck, eds. *Reflexive Modernisation*. London: Polity, 1997.
Glasman, M. 'Society Not State: The Challenge of the Big Society.' *Public Policy Research* 17, no. 2 (2010): 59–63.
Global Reporting Initiative. 'Global Reporting Initiative: Home.' Global Reporting Initiative, http://www.globalreporting.org.
—. 'Sustainability Reporting Guidelines: Version 3.0.' New York: Collaborating Centre of the United Nations Environment Programme, 2006.
Goodin, R.E. *Green Political Theory*. Cambridge: Polity, 1992.

Goodman, M.K. 'The Mirror of Consumption: Celebritization, Developmental Consumption and the Shifting Politics of Fair Trade.' *Geoforum* 41 (2010): 104–16.

Goodman, M.K., D. Maye, and L. Holloway. 'Ethical Foodscapes?: Premises, Promises, Possibilities.' *Environment and Planning A* 42, no. 8 (2010): 1782–96.

Gordon, J. 'Companies Need a Conscience: Hewson.' *The Age*, 14 October 2003, 4.

Gorz, A. *Reclaiming Work: Beyond the Wage-Based Society*. Cambridge: Polity, 1999 [1997].

Gow, D. 'Osborne Wins Wide Support in Fight against Tobin Tax.' *The Guardian*, 8 November 2011.

Gray-Barkan, T. 'Wal-Mart and Beyond: The Battle for Good Jobs and Strong Communities in Urban America.' Los Angeles: Los Angeles Alliance for a New Economy/Partnership for Working Families, 2007.

Grayson, I. 'Green Consumerism.' *Chain Reaction*, Autumn 1989, 27–9.

Grayson, R.S. 'Localism the American Way.' *Public Policy Research* 17, no. 2 (2010): 75–9.

GRI, UNEP, KPMG, UCGA. 'Carrots and Sticks – Promoting Transparency and Sustainability: An Update on Trends in Mandatory and Voluntary Sustainability Reporting.' Nairobi: Global Reporting Initiative, United Nations Environment Programme, KPMG, Unit for Corporate Governance in Africa, 2009.

Gunder, M. 'Sustainability: Planning's Saving Grace or Road to Perdition?.' *Planning Education and Research* 26 (2006): 208–21.

Gupta, J. 'North-South Aspects of the Climate Change Issue: Towards a Negotiating Theory and Strategy for Developing Countries.' *International Journal of Sustainable Development* 3, no. 2 (2000): 115–35.

Habermas, J. *Legitimation Crisis*. Translated by T. McCarthy. Boston: Beacon Press, 1975 [1973].

——. *Lifeworld and System: A Critique of Functionalist Reason*. Translated by T. McCarthy. The Theory of Communicative Action: Vol.2. Boston: Beacon Press, 1987 [1985].

——. *The Structural Transformation of the Public Sphere: An Inquiry into a Category of Bourgeois Society*. Translated by T. Burger and F. Lawrence. Boston Polity Press/ Massachusetts Institute of Technology, 1989 [1962].

——. *Toward a Rational Society: Student Protest, Science, and Politics*. Translated by J.J. Shapiro. London: Heinemann, 1971 [1968].

Haddock, Fiona. 'Socially Responsible Companies: Corporate Angels.' *Global Finance* 13, no. 12 (December 1999): 24–7.

Hagerty, M.R. 'Unifying Livability and Comparison Theory: Cross-National Time-Series Analysis of Life-Satisfaction.' *Social Indicators Research* 47, no. 1 (1999): 343–56.

Hajer, M. 'Ecological Modernisation as Cultural Politics.' In *Risk, Environment, Modernity*, edited by S. Lash and B. Szerszynski. 246–68. London: Sage, 1996.

——. 'A Frame in the Fields: Policymaking and the Reinvention of Politics.' In *Deliberative Policy Analysis: Understanding Governance in the Network Society*, edited by M. Hajer and H. Wagenaar. 88–112. Cambridge: Cambridge University Press, 2003.

——. 'Policy without Polity? Policy Analysis and the Institutional Void.' *Policy Sciences* 36 (2003): 175–95.

——. *The Politics of Environmental Discourse: Ecological Modernisation and the Policy Process*. Oxford: Clarendon Press, 1995.

Hajer, M., and H. Wagenaar, eds. *Deliberative Policy Analysis: Understanding Governance in the Network Society.* edited by R.E. Goodin, Theories of Institutional Design. Cambridge: Cambridge University Press, 2003.

Hamilton, C. 'What's Left? The Death of Social Democracy.' *Quarterly Essay*, no. 21 (March 2006): 1–69.

Handler, J. 'Reforming/Deforming Welfare.' *New Left Review* II, no. 4 (July–August 2000): 114–36.

Harris, P. 'From Relief to Mutual Obligation: Welfare Rationalities and Unemployment in 20th-Century Australia.' *Journal of Sociology* 37, no. 1 (2001): 5–26.

Harris, P.G. *World Ethics and Climate Change.* Edinburgh: Edinburgh University Press, 2010.

Harvey, D. *The Condition of Postmodernity: An Enquiry into the Origins of Cultural Change.* Oxford: Blackwell Books, 1990.

Hayward, T. 'Anthropocentrism: A Misunderstood Problem.' *Environmental Values* 6, no. 1 (1997): 49–63.

——. *Constitutional Environmental Rights.* Oxford: Oxford University Press, 2005.

——. 'Ecological Citizenship: Justice, Rights and the Virtue of Resourcefulness.' *Environmental Politics* 15, no. 3 (2006): 435–46.

——. 'Global Justice and the Distribution of Natural Resources.' *Political Studies* 54, no. 2 (2006): 349–69.

——. *Political Theory and Ecological Values.* Cambridge: Polity, 1998.

Heath, J. *Efficient Society: Why Canada Is as Close to Utopia as It Gets.* Toronto: Penguin Global, 2005.

Heath, J., and A. Potter. *Rebel Sell: Why the Culture Can't Be Jammed.* New York: HarperBusiness, 2004.

Heelas, P. 'Work Ethics, Soft Capitalism and the "Turn to Life".' In *Cultural Economy*, edited by P. du Gay and M. Pryke. 78–96. London: Sage Publications, 2002.

Hempel, L.C. 'Conceptual and Analytical Challenges in Building Sustainable Communities.' In *Toward Sustainable Communities: Transition and Transformations in Environmental Policy*, edited by D.A. Mazmanian and M.E. Kraft. 43–74. Cambridge: MIT Press, 1999.

Hendriks, C.M. 'Deliberative Governance in the Context of Power.' *Policy and Society* 28 (2009): 173–84.

——. 'Institutions of Deliberative Democratic Process and Interest Groups: Roles, Tensions and Incentives.' *Australian Journal of Public Administration* 61, no. 1 (2002): 64–75.

Hendriks, C.M., J.S. Dryzek, and C. Hunold. 'Turning up the Heat: Partisanship in Deliberative Innovation.' *Political Studies* 55 (2007): 362–83.

Hess, D.J. *Localist Movements in a Global Economy: Sustainability, Justice, and Urban Development in the United States.* Cambridge: MIT Press, 2009.

Hill, R. 'Introducing the Triple Bottom-Line.' KPMG Global, http://www.kpmg.com/Global/en/IssuesAndInsights/ArticlesPublications/Press-releases/Pages/Press-release-Introducing-the-triple-bottom-line-1-Mar-2010.aspx.

Hill, R.P., and D.L. Stephens. 'The Compassionate Organisation in the 21st Century.' *Organizational Dynamics* 32, no. 4 (2003): 331–41.

Hilton, M. 'Consumers and the State since the Second World War.' *ANNALS* 611 (2007): 66–81.

Hirschman, A.O. *The Rhetoric of Reaction: Perversity, Futility, Jeopardy*. Cambridge: Belknap Press, 1991.

Hobson, K. 'Bins, Bulbs and Shower Timers: On the "Techno-Ethics" of Sustainable Living.' *Ethics, Place & Environment* 9, no. 3 (2006): 317–36.

——. 'Thinking Habits into Action: The Role of Knowledge and Process in Questioning Household Consumption Practices.' *Local Environment* 8, no. 1 (2003): 95–112.

Hochschild, A. *The Time Bind: When Work Becomes Home and Home Becomes Work*. New York: Metropolitan Books, 1997.

Holden, M. 'Community Interests and Indicator System Success.' *Social Indicators Research* 92 (2009): 429–48.

——. 'Revisiting the Local Impact of Community Indicators Projects: Sustainable Seattle as a Prophet in Its Own Land.' *Applied Research in Quality of Life* 1 (2006): 253–77.

——. 'The Tough-Minded and the Tender-Minded: A Pragmatic Turn for Urban Sustainable Development Policy.' *Planning Theory and Practice* 9, no. 4 (2008): 475–96.

——. 'Urban Indicators and the Integrative Ideals of Cities.' *Cities* 23, no. 3 (2006): 170–83.

Holden, M., and C. Mochrie. 'Counting on Vancouver: Our View of the Region. Inaugural Report of the Regional Vancouver Urban Observatory.' Vancouver: Regional Vancouver Urban Observatory/Simon Fraser University Urban Studies Program, 2006.

——. 'The Regional Vancouver Urban Observatory (RVU): Counting on Vancouver, "Our View" of the Region.' *Revista Internacional de Sostenabilidad, Tecnologia y Humanismo* 1 (2006): 137–244.

Holman, N. 'Incorporating Local Sustainability Indicators into Structures of Local Governance: A Review of the Literature.' *Local Environment* 14, no. 4 (2009): 365–75.

Honderich, T. *Conservatism: Burke, Nozick, Bush, Blair?* Rev. ed. London: Pluto Press, 2005 [1989].

Honneth, A. 'Organized Self-Realization: Some Paradoxes of Individualization.' *European Journal of Social Theory* 7, no. 4 (2004): 463–78.

Humm, J., K. Jones, and G. Chanan. 'Testing Indicators of Community Involvement: Final Report.' London: Community Development Foundation, 2005.

ILO. 'ILO Declaration on Social Justice for a Fair Globalization.' Geneva: International Labour Organization, 2008.

——. 'Social Protection Floor for a Fair and Inclusive Globalization: Report of the Advisory Group.' Geneva: International Labour Organization with World Health Organization, 2011.

Inglehart, R. *Culture Shift in Advanced Industrial Societies*. Princeton: Princeton University Press, 1990.

——. *Modernization and Postmodernization*. Princeton: Princeton University Press, 1997.

——. *The Silent Revolution: Changing Values and Political Styles among Western Publics*. Princeton: Princeton University Press, 1977.

International Council for Local Environmental Initiatives. 'Home Page.' http://www.iclei.org.

International Standards Organization. 'Guidance on Social Responsibility.' ISO, http://www.iso.org/iso/about.htm.
Isin, E.F. 'Citizenship, Class and the Global City.' *Citizenship Studies* 3, no. 2 (1999): 267–83.
——. 'The Neurotic Citizen.' *Citizenship Studies* 8, no. 3 (2004): 217–35.
Isin, E.F., and P.K. Wood. *Citizenship and Identity*. London: Sage, 1999.
Jackson, A. 'Doing the Right Thing: Who Cares, Wins.' *Sydney Morning Herald*, 7 June 2003, 50.
Jackson, K.T. *Crabgrass Frontier: The Suburbanization of the United States*. Oxford: Oxford University Press, 1985.
James, P. *Globalism, Nationalism, Tribalism: Bringing Theory Back In*. London: Sage Publications, 2006.
——. *Nation Formation: Towards a Theory of Abstract Community*. London: Sage Publications, 1996.
Jenkins, S. 'All This Drivel Does Is Bring Basra to Our Doorsteps.' *The Guardian* (11 January 2006): http://www.guardian.co.uk/Columnists/Column/0,5673,1683657,00.html.
——. *Thatcher and Sons: A Revolution in Three Acts*. London: Penguin Books, 2006.
Jessop, B. 'The Changing Governance of Welfare: Recent Trends in Its Primary Functions, Scale and Modes of Coordination.' *Social Policy and Administration* 33, no. 4 (1999): 348–59.
——. 'Liberalism, Neoliberalism and Urban Governance: A State-Theoretical Perspective.' *Antipode* 34, no. 3 (2002): 452–72.
Joas, H. *The Creativity of Action*. Translated by J. Gaines and P. Keast. Chicago: University of Chicago Press, 1996 [1992].
John, D. 'Civic Environmentalism.' In *Environmental Governance Reconsidered: Challenges, Choices, Opportunities*, edited by R.F. Durant, D.J. Fiorino, and R. O'Leary. 219–54. Cambridge: MIT Press, 2004.
——. 'Civic Environmentalism.' *Issues in Science and Technology* 10, no. 4 (1994): 30–4.
John, D., and M. Mlay. 'Community-Based Environmental Protection: Encouraging Civic Environmentalism.' In *Better Environmental Decisions: Strategies for Governments, Businesses and Communities*, edited by K. Sexton, A.A. Marcus, W. Easter, and T.D. Burkhardt. 353–76. Washington: Island Press, 1999.
Jordan, A., R. Wurzel, and A. Zito. 'The Rise of 'New' Policy Instruments in Comparative Perspective: Has Governance Eclipsed Government?.' *Political Studies* 53, no. 3 (2005): 477–96.
Juniper, C., and M. Moore. 'Synergies and Best Practices of Corporate Partnerships for Sustainability.' *Corporate Environmental Strategy* 9, no. 3 (2002): 267–76.
Kasperson, R.E. 'Rerouting the Stakeholder Express.' *Global Environmental Change* 16, no. 4 (2006): 320–2.
Keil, R. '"Common-Sense" Neoliberalism: Progressive Conservative Urbanism in Toronto, Canada.' *Antipode* 34, no. 3 (2002): 578–601.
Kelly, G., D. Kelly, and A. Gamble. *Stakeholder Capitalism*. London: Macmillan, 1997.
Kelly, R. 'Community Values Create Value.' *Business Review Weekly*, 16 January 2003, 447.

Kemple, T.M. 'Spirits of Late Capitalism.' *Theory, Culture & Society* 24, no. 3 (2007): 147–59.

KPMG. 'Beyond Numbers™: How Leading Organisations Link Values with Value to Gain Competitive Advantage.' KPMG, http://www.kpmg.com.

KPMG Climate Change & Sustainability Services. 'KPMG Global: KPMG Climate Change & Sustainability Services.' KPMG Global, http://www.kpmg.com/Global/en/WhatWeDo/Advisory/Risk-Compliance/Internal-Audit/Climate-Change-Sustainability-Services/Pages/Climate-change-Sustainability-Services.aspx.

Krippner, G.R. 'The Financialization of the American Economy.' *Socio-Economic Review*, no. 3 (2005): 173–208.

Kristol, I. 'Keeping up with Ourselves.' In *The End of Ideology*, edited by C.I. Waxman. 18 March 2006. New York: Funk & Wagnalls, 1968.

Lafaye, C., and L. Thévenot. 'Une Justification Écologique?: Conflits Dans L'aménagement De La Nature.' *Revue Française de Sociologie* 34, no. 4 (1993): 495–524.

Lagan, A. 'A Revolution to Find Value in Common Ground.' *Australian Financial Review*, 29 May 2003, 23.

Lamont, M., and L. Thévenot, eds. *Rethinking Comparative Sociology: Repertoires of Evaluation in France and the United States*. edited by J.C. Alexander and S. Seidman, Cambridge Cultural Social Studies. Cambridge: Cambridge University Press, 2000.

Lander, C. *Stuff White People Like: A Definitive Guide to the Unique Taste of Millions*. New York: Random House, 2008.

Lasch, C. *The Culture of Narcissism: American Life in an Age of Diminishing Expectations*. New York: W. W. Norton and Co., 1979.

——. *Haven in a Heartless World: The Family Besieged*. New York: Basic Books, 1977.

Lascoumes, P. *L'eco-Pouvoir, Environnements Et Politiques*. Paris: La Découverte, 1994.

——. 'Les Instruments D'action Publique, Traceurs De Changement: L'exemple Des Transformations De La Politique Française De Lutte Contre La Pollution Atmospherique (1961–2006).' *Politique et Societies* 26, no. 2–3 (2007): 73–89.

Latour, B. 'Love Your Monsters.' In *Love Your Monsters: Postenvironmentalism and the Anthropocene*, edited by M. Shellenberger and T. Nordhaus. 16–23. Washington: Breakthrough Institute, 2011.

——. *Politics of Nature: How to Bring the Sciences into Democracy* Cambridge: Harvard University Press, 2004.

——. *We Have Never Been Modern*. Translated by C. Porter. Cambridge: Harvard University Press, 1993 [1991].

Latta, A. 'Locating Democratic Politics in Ecological Citizenship.' *Environmental Politics* 16, no. 3 (2007): 377–93.

Lazonick, W., and M. O'Sullivan. 'Maximizing Shareholder Value: A New Ideology for Corporate Governance.' *Economy and Society* 29, no. 1 (February 2000): 13–35.

Le Galès, P., and A. Scott. 'Une Revolution Bureaucratique Britannique? Autonomie Sans Controle Ou "Freer markets, More Rules".' *Revue Française de Sociologie* 49, no. 2 (2008): 301–30.

Lears, J. *Fables of Abundance*. New York: Harper Collins Publishing, 1994.
——. 'Reconsidering Abundance: A Plea for Ambiguity.' In *Getting and Spending*, edited by S. Strasser et al. 449–66. Washington: German Historical Institute, 1998.
Lefort, C. 'Outline of the Genesis of Ideology in Modern Societies.' In *The Political Forms of Modern Society: Bureaucracy, Democracy, Totalitarianism*, edited by J.B. Thompson. 181–235. Cambridge: MIT Press, 1986 [1974].
——. *The Political Forms of Modern Society: Bureaucracy, Democracy, Totalitarianism*. Cambridge: MIT Press, 1986.
Lewis, G. 'New Labour and Post-Fordist Ideology: Inherent Contradictions.' *Arena* 23, no. New Series (2005): 19–28.
Lloyd, G. *The Man of Reason: 'Male' and 'Female' in Western Philosophy*. Ideas. edited by J. Ree. London: Methuen, 1984.
Lohmann, L. *Carbon Trading: A Critical Conversation on Climate Change, Privatisation and Power*. Development Dialogue. edited by N. Hällström, O. Nordberg, and R. Österbergh. Vol. 48. Uppsala: Dag Hammarskjöld Centre/Corner House, 2006.
Luhmann, N. 'Limits of Steering.' *Theory, Culture & Society* 14, no. 1 (1997): 41–57.
——. *Observations on Modernity*. Stanford: Stanford University Press, 1998.
——. *Risk*. New York: Aldine De Gruyter, 1993.
Lukes, S. 'Epilogue: The Grand Dichotomy of the Twentieth Century.' In *The Cambridge History of Twentieth-Century Political Thought*, edited by T. Ball and R. Bellamy. 602–26. Cambridge: Cambridge University Press, 2006.
Luttwak, E.N. *Turbo-Capitalism: Winners and Losers in the Global Economy*. London: Weidenfeld & Nicolson/HarperCollins, 2000.
Lyon, T.P., and J.W. Maxwell. 'Corporate Social Responsibility and the Environment: A Theoretical Perspective.' *Review of Environmental Economics and Policy* 2, no. 2 (2008): 240–60.
Lyotard, J-F. *The Condition of Postmodernity: A Report on Knowledge*. Translated by G. Bennington and D. Massumi. 9th ed. Minneapolis: University of Minnesota Press, 1993 [1979].
Maffesoli, M. 'The Return of Dionysus.' In *Constructing the New Consumer Society*, edited by P. Sulkunen, J. Holmwod, H. Radner, and G. Schulze. 21–37. New York: St Martin's Press, 1997.
——. *The Time of the Tribes: The Decline of Individualism in Mass Society*. Theory, Culture & Society. edited by M. Featherstone London: Sage Books, 1996 [1988].
Manteaw, B. 'From Tokenism to Social Justice: Rethinking the Bottom Line for Sustainable Community Development.' *Community Development Journal* 43, no. 4 (2007): 428–43.
Marsh Mercer Kroll. 'Quality of Living – Survey Highlights.' Marsh Mercer Kroll, http://www.mercer.com/qualityofliving.
Marshall, T.H. 'Citizenship and Social Class.' In *Citizenship and Social Class*, edited by T.H. Marshall and T. Bottomore. Part I. London: Pluto Press, 1992.
——. 'Citizenship and Social Class.' In *The Citizenship Debates: A Reader*, edited by G. Shafir. 93–111. Minneapolis: University of Minnesota Press, 1998 [1950].
——. *Class, Citizenship, and Social Development*. New York: Free Press, 1965.

——. *The Right to Welfare and Other Essays*. London: Heinemann, 1981.

Martinez-Alier, J. *The Environmentalism of the Poor: A Study of Ecological Conflicts and Valuation*. Cheltenham: Edward Elgar, 2002.

May, D. 'Uses of Community Accounts.' 14. St Johns: Newfoundland and Labrador Statistics Agency/ Memorial University of Newfoundland, 2005.

Mayo, E. 'Dreams of a Social Economy.' *Public Policy Research* 17, no. 2 (2010): 64–6.

McAlpine, P., and A. Birnie. 'Is There a Correct Way of Establishing Sustainability Indicators? The Case of Sustainability Indicator Development on the Island of Guernsey.' *Local Environment* 10, no. 3 (2005): 243–57.

McKinlay, P. 'The Challenge of Democratic Participation in the Community Development Process.' *Community Development Journal* 41, no. 4 (2006): 492–505.

Meadowcroft, J. 'From Welfare State to Ecostate.' In *The State and the Global Ecological Crisis*, edited by J. Barry and R. Eckersley. 3–24. Cambridge: MIT Press, 2005.

Melucci, A. *The Playing Self: Person and Meaning in the Planetary Society*. Cambridge Cultural Social Studies. edited by J.C. Alexander and S. Seidman. Cambridge: Cambridge University Press, 1996.

Meyer, J.M. *Political Nature: Environmentalism and the Interpretation of Western Thought*. Cambridge: MIT Press, 2001.

Miller, C. 'New Civic Epistemologies of Quantification: Making Sense of Indicators of Local and Global Sustainability.' *Science, Technology and Human Values* 30, no. 3 (2005): 403–32.

Miller, P., and N. Rose. *Governing the Present: Administering Economic, Social and Personal Life*. Cambridge: Polity, 2008.

Mol, A.P.J. *Globalization and Environmental Reform: The Ecological Modernisation of the Global Economy*. Cambridge: MIT Press, 2001.

Mol, A.P.J., and G. Spaargaren. 'Environment, Modernity and the Risk-Society.' *International Sociology* 8, no. 4 (1993): 431–59.

Moldan, B., and A. Lyon Dahl. 'Challenges to Sustainability Indicators.' In *Scope 67: Sustainability Indicators a Scientific Assessment*, edited by T. Hak, B. Moldan, and A. Lyon Dahl. 1–24. Washington: Island Press, 2007.

Moldan, B., J.W.B. Stewart, and V. Plocq-Fichelet. 'Preface.' In *Scope 67: Sustainability Indicators a Scientific Assessment*, edited by T. Hak, B. Moldan, and A. Lyon Dahl. xxiii–xxv. Washington: Island Press, 2007.

Moody, M., and L. Thévenot. 'Comparing Models of Strategy, Interests and the Public Good in French and American Environmental Disputes.' In *Rethinking Comparative Cultural Sociology*, edited by L. Thévenot and M. Lamont. 273–306. Cambridge: Cambridge University Press, 2001.

Mort, F. 'Competing Domains: Democratic Subjects and Consuming Subjects in Britain and the United States since 1945.' In *The Making of the Consumer: Knowledge, Power and Identity in the Modern World*, edited by F. Trentmann. Oxford: Berg Publishing, 2006.

Mukerji, C. *From Graven Images: Patterns of Modern Materialism*. New York: Columbia University Press, 1983.

Müller, J-W. 'Comprehending Conservatism: A New Framework for Analysis.' *Journal of Political Ideologies* 11, no. 3 (2006): 359–65.

Murray, C. 'Guaranteed Income as a Replacement for the Welfare State [Foundation for Law, Justice and Society Policy Brief, Oxford (http://Www.Fljs.Org)%5D.' *Basic Income Studies* 3, no. 2 (2007): 1–12.
National Endowment for the Arts. *The NEA 1965–2000: A Brief Chronology of Federal Support for the Arts*. Washington DC: NEA, 2000.
Newell, P., and M. Paterson. *Climate Capitalism: Global Warming and the Transformation of the Global Economy*. Cambridge: Cambridge University Press, 2010.
Nussbaum, M. *Creating Capabilities: The Human Development Approach*. Cambridge: Harvard University Press/Belknap, 2011.
O'Connor, J. *Accumulation Crisis*. Oxford: Basil Blackwell, 1984.
——. *The Fiscal Crisis of the State*. New York: Transaction Books, 2009 [1973].
Obama, B. 'Inaugural Address.' In, (2009). http://obamaspeeches.com.
OECD. 'Better Life Index.' Organization for Economic Co-operation and Development, http://www.oecdbetterlifeindex.org/.
——. 'Working Together Towards Sustainable Development: The OECD Experience.' Paris: Organization for Economic Co-operation and Development, 2002.
Offe, C. *Contradictions of the Welfare State*. Cambridge: MIT Press, 1984.
Offer, A. *The Challenge of Affluence: Self-Control and Well-Being in the United States and Britain since 1950*. Oxford: Oxford University Press, 2006.
Olsen, J., M. Koss, and D. Hough. 'From Pariahs to Players? Left Parties in National Governments.' In *Left Parties in National Governments*, edited by J. Olsen, M. Koss, and D. Hough. London: Palgrave Macmillan, 2010.
Paehlke, R.C. *Environmentalism and the Future of Progressive Politics*. New Haven: Yale University Press, 1989.
Parry, M.L., O.F. Canziani, J.P. Palutikof, P.J. van der Linden, and C.E. Hanson, eds. *Fourth Assessment Report of the Intergovernmental Panel on Climate Change*. Cambridge: Cambridge University Press, 2007.
Patomakki, H. *Democratising Globalisation: The Leverage of the Tobin Tax*. London: Zed Books, 2001.
Paz-Fuchs, A. 'Equality and Personal Responsibility in the New Social Contract.' In *Report and Analysis of the 5th Workshop of the Social Contract Revisited*. Oxford University: The Foundation for Law, Justice and Society, 2009.
Peck, J. 'Struggling with the Creative Class.' *International Journal of Urban and Regional Research* 29, no. 4 (2005): 740–70.
Peel, M. *The Lowest Rung: Voices of Australia's Poverty*. Cambridge: Cambridge University Press, 2004.
Perceval, C. 'Book Review: *The Chrysalis Economy* by Elkington, J.' *Business & Society* 41, no. 2 (June 2002): 258–61.
Peters, T., and R.H. Waterman. *In Search of Excellence*. New York: Harper & Row Publishers, 1982.
Petersen, L.K. 'Changing Public Discourse on the Environment: Danish Media Coverage of the Rio and Johannesburg UN Summits.' *Environmental Politics* 16, no. 2 (2007): 206–30.
Petit, P. *Civic Republicanism*. Oxford: Oxford University Press, 1997.
Pierre, J., ed. *Debating Governance*. Oxford: Oxford University Press, 2000.
Pierre, J., and B.G. Peters. *Governance, Policy and the State*. London: Macmillan, 2000.

Pierson, C. 'Lost Property: What the Third Way Lacks.' *Journal of Political Ideologies* 10, no. 2 (2005): 145–63.
Pimlott, D. 'Tobin Tax Remains Treasury Ambition.' *Financial Times*, 11 December 2009.
Plumwood, V. 'Inequality, Ecojustice and Ecological Rationality.' *Ecotheology: Journal for the Study of Religion, Nature and Culture* 5/6 (1999): 185–218.
——. 'Nature, Self and Gender: Feminism, Environmental Philosophy and the Critique of Rationalism.' *Hypatia* 6, no. 1 (1991): 4–32.
Polanyi, K. *The Great Transformation*. New York: Beacon Books, 2001 [1944].
Pollock, R.M., and G.S. Whitelaw. 'Community-Based Monitoring in Support of Local Sustainability.' *Local Environment* 10, no. 3 (2005): 211–28.
Porter, I. 'General Electric Warms up a Slice Nuclear Pie.' *The Age*, 26 May 2006, 2.
Portney, P.R. 'The (Not So) New Corporate Social Responsibility: An Empirical Perspective.' *Review of Environmental Economics and Policy* 2, no. 2 (2008): 261–75.
Possomaï, A. 'Cultural Consumption of History and Popular Culture in Alternative Spiritualities.' *Journal of Consumer Culture* 2, no. 2 (1 July 2002).
——. 'Secrecy and Consumer Culture: An Exploration of Esotericism in Contemporary Western Society Using the Work of Simmel and Baudrillard.' *Australian Religious Studies Review* 15, no. 1 (2001): 44–50.
Prabhakar, R. 'Stakeholding: Does It Possess a Stable Core?.' *Journal of Political Ideologies* 8, no. 3 (2003): 347–63.
Prime Minister's Office. 'Government Launches "Big Society" Programme.' Office of the Prime Minister, http://www.number10.gov.uk/news/topstorynews/2010/05/big-society-50248.
Princen, T. *The Logic of Sufficiency*. Cambridge: MIT Press, 2005.
——. 'Principles for Sustainability: From Cooperation and Efficiency to Sufficiency.' *Global Environmental Politics* 3, no. 1 (2003): 33–50.
Raco, M. 'Securing Sustainable Communities: Citizenship, Safety and Sustainability in the New Urban Planning.' *European Urban and Regional Studies* 14, no. 4 (2007): 305–20.
Rayner, M., R. Harrison, and S. Irving. 'Ethical Consumerism: Democracy through the Wallet.' *Journal of Research for Consumers* 1, no. 3 (June 2002): 1–12.
Reich, C.A. *The Greening of America*. London: Penguin Books, 1995 [1970].
Rhodes, R.A.W. *Understanding Governance*. Buckingham: Open University Press, 1997.
Ritzer, G. *Enchanting a Disenchanting World: Revolutionising the Means of Consumption*. Thousand Oaks: Pine Forge Press, 1999.
Roberts, D. 'The Full Text of Obama's Energy Remarks.' Grist Magazine inc., http://www.grist.org/article/obama_factsheet.
Rosanvallon, P. *Democratic Legitimacy: Impartiality, Reflexivity, Proximity*. Translated by A. Goldhammer. Princeton: Princeton University Press, 2011 [2008].
——. *The New Social Question: Rethinking the Welfare State*. Princeton: Princeton University Press, 2000 [1995].
Rose, N. *Powers of Freedom: Reframing Political Thought*. Cambridge: Cambridge University Press, 1999.

Rosenau, J.N. 'Governance, Order and Change in World Politics.' In *Governance without Government*, edited by J.N. Rosenau and E-O. Czemoiel. 1–29. Cambridge: Cambridge University Press, 1992.

Ross, A. *No Collar: The Humane Workplace and Its Hidden Costs*. New York: Basic Books, 2002.

Roszak, T. *The Making of a Counterculture: Reflections on the Technocratic Society & Its Youthful Opposition*. 2nd ed. Berkeley: University of California Press, 1995 [1968].

Rowley, T.J., and S. Berman. 'A Brand New Brand of Corporate Social Performance.' *Business & Society* 39, no. 4 (2000): 397–412.

Rutland, T., and A. Aylett. 'The Work of Policy: Actor Networks, Governmentality, and Local Action on Climate Change in Portland, Oregon.' *Environment and Planning D: Society and Space* 26 (2008): 627–46.

Rydin, Y. 'Indicators as a Governmental Technology? The Lessons of Community-Based Sustainability Indicator Projects.' *Environment and Planning D: Society and Space* 25 (2007): 610–24.

Rydin, Y., N. Holman, and E. Wolff. 'Editorial: Local Sustainability Indicators.' *Local Environment* 8, no. 6 (2003): 581–89.

Sainsbury, R. '21st Century Welfare – Getting Closer to Radial Benefit Reform.' *Public Policy Research* 17, no. 2 (2010): 102–7.

Salmon, A. *La Tentation Éthique Du Capitalisme*. Paris: La Découverte, 2007.

——. *Moraliser Le Capitalisme*. Paris: CNRS Éditions, 2009.

Sandercock, L. 'The 'Wal-Marting' of the World? The Neoliberal City in North America.' *City & Community* 9, no. 1 (2005): 3–8.

Sandler, R., and P.C. Pezzullo, eds. *Environmental Justice and Environmentalism: The Social Justice Challenge to the Environmental Movement*. Cambridge: MIT Press, 2007.

Sassen, S. *Globalization and Its Discontents: Essays on the New Mobility of People and Money*. New York: New Press, 1998.

Sassoon, D. *One Hundred Years of Socialism: The West European Left in the Twentieth Century*. London: I.B. Tauris, 2010.

Scerri, A. 'Accounting for Sustainability: Implementing a Residential Emissions Reduction Strategy Using an Approach That Combines Qualitative and Quantitative "Indicators" of Sustainability.' *Management of Environmental Quality* 21, no. 10 (2010): 122–35.

——. 'Livability Index.' In *Encyclopedia of Quality of Life Research*, edited by A. Michalos. Heidelberg: Springer, 2013, in press.

——. 'On Throwing out the Baby with the Bathwater.' *Arena Magazine* no. 82 (April–May 2006): 53–4.

——. 'Paradoxes of Increased Individuation and Public Awareness of Environmental Issues.' *Environmental Politics* 18, no. 4 (2009): 467–85.

——. 'Rethinking Responsibility? Household Sustainability in the Stakeholder Society.' In *Material Geographies of Household Sustainability*, edited by R. Lane and A. Gorman-Murray. 175–92. London: Ashgate, 2011.

——. 'Review Article: The New Extremism in 21st Century Britain.' *Politics, Religion & Ideology* 12, no. 1 (2011): 119–20.

——. 'Triple Bottom-Line Capitalism and the 3rd Place.' *Arena Journal* New Series, no. 20 (2003): 56–67.

Schlosberg, D. *Defining Environmental Justice*. Oxford: Oxford University Press, 2007.
——. *Environmental Justice and the New Pluralism: The Challenge of Difference for Environmentalism*. Oxford: Oxford University Press, 1999.
Schlosberg, D., and S. Rinfret. 'Ecological Modernisation, American Style.' *Environmental Politics* 17, no. 2 (2008): 254–75.
Schor, J.B. *The Overworked American: The Unexpected Decline of Leisure*. New York: Basic Books/HarperCollins Publishing, 1991.
Schudson, M. 'The Troubling Equivalence of Citizen and Consumer.' *Annals of the American Academy of Political and Social Sciences* 608 (2006): 193–204.
Schulze, G. *The Experience Society*. London: Sage Publications, 1995.
——. 'From Situations to Subjects: Moral Discourse in Transition.' In *Constructing the New Consumer Society*, edited by P. Sulkunen, J. Holmwod, H. Radner, and G. Schulze. 38–57. New York: St Martin's Press, 1997.
Scott, M. 'Reflections on the Big Society.' *Community Development Journal* 46, no. 1 (2010): 132–7.
Scruton, R. 'Conservatism.' In *Political Theory and the Ecological Challenge*, edited by A. Dobson and R. Eckersley. 7–19. Cambridge: Cambridge University Press, 2006.
Sen, A. *The Idea of Justice*. Cambridge: Harvard University Press, 2009.
——. *On Ethics and Economics*. Oxford: Blackwell, 1988.
Sennett, R. *The Culture of the New Capitalism*. London: Yale University Press, 2005.
Seyfang, G. 'Shopping for Sustainability: Can Sustainable Consumption Promote Ecological Citizenship.' *Environmental Politics* 14, no. 2 (April 2005): 290–306.
Seyfang, G., and A. Smith. 'Grassroots Innovation for Sustainable Development: Towards a New Research and Policy Agenda.' *Environmental Politics* 16, no. 4 (August 2007): 584–603.
Shellenberger, M., and T. Nordhaus. *The Death of Environmentalism: Global Warming Politics in a Postenvironmental World*. Washington: Environmental Grantmakers Association/The Breakthrough Institute, 2004.
——. 'Evolve: The Case for Modernization as the Road to Salvation.' In *Love Your Monsters: Postenvironmentalism and the Anthropocene*, edited by M. Shellenberger and T. Nordhaus. 8–15. Washington: Breakthrough Institute, 2011.
Sklair, L. 'Sociology of the Global System.' In *The Globalization Reader*, edited by F.J. Lechner and J. Boli. 70–6. Oxford: Blackwell Publishing, 2005 [2002].
Skocpol, T. *States and Social Revolutions*. Cambridge: Cambridge University Press, 1979.
Smith, D. 'Social Exclusion, the Third Way and the Reserve Army of Labour.' *Sociology* 41, no. 2 (2007): 365–72.
Smith, G. 'Green Citizenship and the Social Economy.' *Environmental Politics* 14, no. 2 (2005): 273–89.
Soederberg, S. 'Taming Corporations or Buttressing Market-Led Development? A Critical Assessment of the Global Compact.' *Globalizations* 4, no. 4 (2007): 500–13.
Soper, K. 'Alternative Hedonism, Cultural Theory and the Role of Aesthetic Revisioning.' *Cultural Studies* 22, no. 5 (2008): 567–87.

——. 'Rethinking the "Good Life": The Consumer as Citizen.' *Capitalism, Nature, Socialism* 15, no. 3 (2004): 111–16.

Spaargaren, G., and A.P.J. Mol. 'Greening Global Consumption: Redefining Politics and Authority.' *Global Environmental Change* 18, no. 2 (2008): 350–9.

Spahn, P.B. 'International Financial Flows and Transactions Taxes: Survey and Opinions.' Washington, DC: International Monetary Fund, 1995.

Spratt, S. 'A Sterling Solution.' London: Stamp out Poverty/Intelligence Capital Limited, 2006.

Steger, M.. *The Rise of the Global Imaginary*. Oxford: Oxford University Press, 2008.

Stoney, C., and D. Winstanley. 'Stakeholding: Confusion or Utopia? Mapping the Conceptual Terrain.' *Journal of Management Studies* 38, no. 5 (2001): 603–26.

Summerville, J.A., B.A. Adkins, and G. Kendall. 'Community Participation, Rights, and Responsibilities: The Governmentality of Sustainable Development Policy in Australia.' *Environment and Planning C: Government and Policy* 26 (2008): 696–711.

Supiot, A. *Au-Delà De L'emploi: Transformations Du Travail Et Devenir Du Droit Du Travail En Europe. Rapport Pour La Commission Des Communautés Européennes*. Paris: Flammarion, 1999.

——. *Homo Juridicus: On the Anthropological Function of the Law*. Translated by S. Brown. London: Verso, 2007 [2005].

——. 'Law and Labour: A World Market of Norms?.' *New Left Review* I, no. 39 (May–June 2006): 109–21.

Susen, S. 'The Transformation of Citizenship in Complex Societies.' *Classical Sociology* 10, no. 3 (2010): 259–85.

Sustainable Seattle. 'Indicators Principles.' Sustainable Seattle, http://www.sustainableseattle.org.

——. 'Sustainable Seattle and Indicators.' King County, http://www.sustainableseattle.org/Programs/RegionalIndicators/index_html.

Swarts, H., and I. Bogdan Vasi. 'Which U.S. Cities Adopt Living Wage Ordinances? Predictors of Adoption of a New Labor Tactic.' *Urban Affairs Review* 47, no. 6 (2011): 743–74.

Swyngedouw, E. 'Impossible "Sustainability" and the Postpolitical Condition.' In *The Sustainable Development Paradox: Urban Political Economy in the United States and Europe*, edited by R. Kreuger and D. Gibbs. 13–40. New York: The Guilford Press, 2007.

Taylor, C. *The Ethics of Authenticity*. Cambridge: Harvard University Press, 1991.

——. *Sources of the Self: The Making of Modern Identity*. Cambridge: Harvard University Press, 1989.

Taylor, M. 'Community Participation in the Real World: Opportunities and Pitfalls in New Governance Spaces.' *Urban Studies* 44, no. 2 (2009): 297–317.

TEEB. 'The Economics of Ecosystems and Biodiversity: Mainstreaming the Economics of Nature: A Synthesis of the Approach, Conclusions and Recommendations of TEEB.' Geneva: The Economics of Ecosystems and Biodiversity / United Nations Environment Programme 2010.

Thévenot, L. 'The Plurality of Cognitive Formats and Engagements: Moving between the Familiar and the Public.' *European Journal of Social Theory* 10, no. 3 (2007): 409–23.

Thévenot, L., M. Moody, and C. Lafaye. 'Forms of Valuing Nature: Arguments and Modes of Justification in French and American Environmental Disputes.' In *Rethinking Comparative Cultural Sociology*, edited by L. Thévenot and M. Lamont. 229–72. Cambridge: Cambridge University Press, 2001.

Thompson, E.P. *The Making of the English Working Class*. London: Penguin Books, 1980 [1963].

Thrift, N. *Knowing Capitalism*. Theory, Culture & Society. edited by Featherstone. M. London: Sage Publications, 2005.

Timmer, V., and N-K Seymoar. 'Vancouver Working Group Discussion Paper.' In *The World Urban Forum 2006*. Vancouver: UN Habitat – International Centre for Sustainable Cities, 2005.

Tobin, J. 'A Proposal for International Monetary Reform.' *Eastern Economic Journal* 4, no. 3–4 (1978): 153–9.

Torgerson, D. 'Democracy through Policy Discourse.' In *Deliberative Policy Analysis: Understanding Governance in the Network Society*, edited by M. Hajer and H. Wagenaar. 113–38. Cambridge: Cambridge University Press, 2003.

Totaro, P. 'Britain Surprises with Carbon Plan.' *The Age*, 21 May 2011.

Trachtenberg, Z. 'Complex Green Citizenship and the Necessity of Judgement.' *Environmental Politics* 19, no. 3 (2010): 339–55.

Trentmann, F. 'Citizenship and Consumption.' *Journal of Consumer Culture* 7, no. 2 (2007): 147–58.

Turner, B.S. 'Citizenship Studies: A General Theory.' *Citizenship Studies* 1, no. 1 (1997): 5–18.

——. 'Contemporary Problems in the Theory of Citizenship.' In *Citizenship and Social Theory*, edited by B.S. Turner. 1–18. London: Sage, 1993.

——. 'The Erosion of Citizenship.' *British Journal of Sociology* 52, no. 2 (2001): 189–209.

——. 'Extended Review: Justification, the City, and Late Capitalism.' *The Sociological Review* 55, no. 2 (2007): 410–14.

——. 'Marshall, Social Rights and English Identity.' *Citizenship Studies* 13, no. 1 (2009): 65–73.

——. 'Outline of a Theory of Citizenship (Originally Published in *Sociology* 24 (2): 189–217).' In *Citizenship: Critical Concepts*, edited by B.S. Turner and P. Hamilton. 199–226. London: Routledge, 1994 [1990].

Turner, B.S., and C. Rojek. *Society and Culture: Principles of Scarcity and Solidarity*. London: Sage, 2001.

Turner, T. 'Shifting the Frame from Nation-State to Global Market.' *Social Analysis* 46, no. 2 (Summer 2002): 56–80.

UNDP. 'Human Development Reports: History of the Human Development Reports.' United Nations Development Programme, http://hdr.undp.org/en/humandev/reports/.

UNEP. 'Global Green New Deal.' Geneva: United Nations Environment Programme, 2009.

——. 'Towards a Green Economy: Pathways to Sustainable Development and Poverty Eradication.' Geneva: United Nations Environment Programme, 2011.

United Nations. 'Agenda 21.' New York/Rio de Janeiro: United Nations Conference on Environment & Development, 1992.

——. 'Agenda 21 Earth Summit: United Nations Program of Action from Rio.' New York: United Nations, 1992.

——. 'General Assembly Resolution 2 Session 55 United Nations Millennium Declaration.' New York: United Nations, 2000.

United Nations Global Compact. 'United Nations Global Compact: About the GC.' UNGC, http://www.unglobalcompact.org.

Urry, J. 'The "System" of Automobility.' *Theory, Culture & Society* 21, no. 4/5 (2004): 25–39.

Valdivielso, J. 'Social Citizenship and the Environment.' *Environmental Politics* 14, no. 2 (2005): 239–54.

Van Parijs, P. *Arguing for Basic Income*. London: Verso, 1992.

van Steenbergen, B. 'Towards a Global Ecological Citizen.' In *The Condition of Citizenship*, edited by B. van Steenbergen. 141–52. London: Sage, 1994.

Vancouver Foundation. 'Vital Signs for Metro Vancouver.' Vancouver Foundation and Metro Vancouver Regional Authority, http://www.vancouverfoundation-vitalsigns.ca/.

Vanderheiden, S. 'Two Conceptions of Sustainability.' *Political Studies* 56, no. 2 (2008): 435–55.

Veenhoven, R. 'Happiness as an Indicator in Social Policy Evaluation: Some Objections Considered.' In *Between Sociology and Sociological Practice*, edited by K. Mesman Schultz et al. 195–206. Nijmegen: Institute for Applied Social Sciences, 1993.

Veenhoven, R., and J. Ehrhardt. 'The Cross-National Pattern of Happiness: Test of Predictions Implied in Three Theories of Happiness.' *Social Indicators Research* 34, no. 1 (1995): 33–68.

Victorian Local Government Association. 'Liveable & Just Toolkit: Connecting Communities, Strengthening Democracy.' VLGA and the McCaughey Centre at the University of Melbourne, http://www.vlga.org.au/Resources/Liveable_Just_Toolkit.aspx.

Vogel, D. *The Market for Virtue: The Potential and Limits of Corporate Social Responsibility*. Washington, DC: The Brooking Institution, 2005.

Wagner, P. 'After *Justification:* Repertoires of Justification and the Sociology of Modernity.' *European Journal of Social Theory* 2, no. 3 (1999): 341–57.

——. *Modernity as Experience and Interpretation: A New Sociology of Modernity*. Cambridge: Polity Press, 2008.

Wakim, N. 'Les Altermondialistes D'attac "Satisfaits" Mais "Pas Dupes".' *Le Monde*, 16 August 2011.

Weale, A. 'Governance, Government and the Pursuit of Sustainability.' In *Governing Sustainability*, edited by W.N. Adger and A. Jordan. 55–75. Cambridge: Cambridge University Press, 2009.

Weber, M. *From Max Weber: Essays in Sociology*. Translated by H.H. Gerth and C.W. Mills. New York: Oxford University Press, 1958 [1946].

White, S. 'Basic Income Versus Basic Capital: Can We Resolve the Differences?.' *Policy and Politics* 39, no. 167–81 (2011).

——. 'Markets, Time and Citizenship.' *Renewal* 12, no. 1 (2004): 50–63.

——. 'The Republican Critique of Capitalism.' *Critical Review of International Social and Political Philosophy* 14, no. 5 (2011): 561–79.

Wilkinson, R.G. *Poverty and Progress: An Ecological Model of Economic Development*. London: Methuen Books, 1973.

Williams, R. *Problems in Materialism and Culture*. London: Verso Books, 1989.

——. *The Sociology of Culture*. Chicago: University of Chicago Press, 1995. 1981.

Wiseman, J., W. Heine, A. Langworthy, N. McLean, J. Pyke, H. Raysmith, and M. Salvaris. 'Measuring Wellbeing, Engaging Communities.' VicHealth, Victoria University, http://www.communityindicators.net.au.

Wissenburg, M. *Green Liberalism: The Free and the Green Society.* London: UCL Press, 1998.

——. 'Liberalism.' In *Political Theory and the Ecological Challenge,* edited by A. Dobson and R. Eckersley. 20–34. Cambridge: MIT Press, 2006.

Woolcock, G. 'Measuring Up? Assessing the Liveability of Australian Cities.' In *State of Australian Cities: National Conference,* edited by P. Maginn and R. Jones. Canberra: Promaco Conventions, 2009.

WPCCC-RME. 'World People's Conference on Climate Change and the Rights of Mother Earth: Building the People's World Movement for Mother Earth.' WPCCC-RME, pwccc.wordpress.com.

Young, I.M. 'From Guilt to Solidarity.' *Dissent* 50, no. 2 (2003): 39–44.

——. *Responsibility for Justice.* Oxford: Oxford University Press, 2011.

Zimmerman, B. 'Pragmatism and the Capability Approach: Challenges in Social Theory and Empirical Research.' *European Journal of Social Theory* 9, no. 4 (2006): 467–84.

Zuboff, S. 'Evolving: A Call to Action.' *Fast Company* Social Capitalists Issue, no. 78 (January 2004): 97.

Index